D0871894

Understanding Me

Understanding Me

LECTURES AND INTERVIEWS

MARSHALL McLUHAN

Edited by Stephanie McLuhan and David Staines

With a Foreword by Tom Wolfe

M&S

National Library of Canada Cataloguing in Publication

McLuhan, Marshall, 1911-1980
Understanding Me : lectures and interviews / Marshall McLuhan ; edited by Stephanie McLuhan and David Staines ; introduction by Tom Wolfe.

Includes bibliographical references.
ISBN 0-7710-5545-5

1. McLuhan, Marshall, 1911-1980 – Interviews. 2. Mass media specialists – Canada – Interviews. I. McLuhan, Stephanie. II. Staines, David, 1946- . III. Title.

P92.5.M24A4 2003   302.23'092   C2003 902543-8

We acknowledge the financial support of the Government of Canada through the Book Publishing Industry Development Program and that of the Government of Ontario through the Ontario Media Development Corporation's Ontario Book Initiative. We further acknowledge the support of the Canada Council for the Arts and the Ontario Arts Council for our publishing program.

Typeset in Sabon by M&S, Toronto
Printed and bound in Canada

McClelland & Stewart Ltd.
The Canadian Publishers
481 University Avenue
Toronto, Ontario
M5G 2E9
www.mcclelland.com

1 2 3 4 5   07 06 05 04 03

Contents

Understanding Me

Foreword

by Tom Wolfe

Come with me back to the 1990s::::::and the Silicon
Valley::::::and the Internet euphoria::::::and the two w.w.w.
saintly-souls who first prophesied the coming of the World Wide
Web::::::

It was November of 1999, and I was in Palo Alto,
California, the Silicon Valley's de facto capital. Right here
in the Valley the computer industry had produced fourteen
new billionaires in the preceding twelve months. I saw billionaires every morning at breakfast. *Every* morning; the
Valley's power-breakfast scene was a restaurant called Il
Fornaio, which happened to be on the ground floor of my
hotel, the Garden Court. I loved the show. You couldn't have
kept me away.

The billionaires you couldn't miss. They all came in
wearing tight jeans or khakis, shirts with the sleeves rolled up
and the front unbuttoned down to the navel, revealing skin and
chest hair, if any, and leather boating moccasins without socks,
baring the bony structure of their ankles and metatarsals . . .
even the ones up in their fifties who had wire hair sprouting out
of their ears above lobes that sagged as badly as their shoulders
and backs, which were bent over like the letter *n*. They looked
like well-scrubbed beachcombers. Their clothes were so skimpy,

there was no way they could have been carrying a cellphone or even a beeper, let alone a Palm Pilot, a BlackBerry, a RIM pager, or an HP-19B calculator. Walking behind every billionaire would be an aide-de-camp, probably worth no more than 60 or 70 million, wearing the same costume plus a sport jacket. Why a sport jacket? Why, for pockets in which to carry the cellphone, the beeper, the Palm Pilot, the BlackBerry, the RIM pager, and the HP-19B calculator. Billionaires in baby clothes! You could get high in Il Fornaio on second-hand euphoria.

But much of the sublime lift came from something loftier than overnight IPO billions and the like, something verging on the spiritual. Cyberspace had its visionaries, and they were telling everybody in the Valley that they were doing more than simply developing computers and creating a new wonder medium, the Internet. Far more. The Force was with them. They were spinning a seamless web over all the Earth that would forever render national boundaries and racial divisions meaningless and change, literally transform, the nature of the human beast. And everybody in the Valley believed it and dressed the part. Faithful devotees of the Force didn't go about in dull suits and pale blah shirts with "interesting" Hermès neckties and cap-toed black oxfords with shoelaces, the way the dreary, outmoded Wall Street workaday investment donkeys did back East.

The Web – the W was always capitalized – was the world of the future, namely, the Digital Universe, and the Force had its own evangelical journals. *Upside* magazine's editor, Richard L. Brandt, said (September 1998) he expected "to see the overthrow of the U.S. government in my lifetime," not by revolutionaries or foreign aggressors, however, but by Bill Gates's Microsoft. The software Gates and Microsoft provided for the World Wide Web "will gradually make the U.S. government obsolete." Compared to that, Gates himself was Modesty in sneakers when he wrote that he was part of "an

epochal change" that "will affect the world seismically."
Seismically means like an earthquake. Evolution used to be
measured in units of one hundred thousand years. But
computer scientist Danny Hillis wrote in *Wired* magazine that
thanks to "telephony, computers, and CD-ROMs," today "evolution takes place in microseconds. . . . We're taking off. . . .
We are not evolution's ultimate product. There's something
coming after us, and I imagine it is something wonderful. But
we may never be able to comprehend it, any more than a
caterpillar can imagine turning into a butterfly."

Euphoria, as I say, a Millennial vision – and all of it had
been inspired by a Canadian literary scholar who had died
fifteen years before the Internet existed. His name, unknown
outside of Canada until he published the book *Understanding
Media* in 1964, was Marshall McLuhan. By 1996 the cyber-
faithful were looking to McLuhan's work and prophesies as
the new theory of evolution.

I can't think of another figure who so dominated an entire
field of study in the second half of the twentieth century. At
the turn of the nineteenth century and in the early decades
of the twentieth there was Darwin in biology, Marx in political science, Einstein in physics, and Freud in psychology. Since
then there has been only McLuhan in communications studies
or, to be more accurate, McLuhan and a silent partner. It was
the silent partner who made McLuhanism what it was: a
scientific theory set upon an unseen, unspoken, taboo religious base.

McLuhan had been raised as a Baptist in, to all outward
appearances, a family typical of the settlers of the vast
Canadian West. They were Scotch-Irish Protestants who said
howse and *abowt* for house and about. His father's forbears
were farmers. His father himself was an insurance salesman. But
his mother, Elsie Hall McLuhan, was another story. She was the
cosmopolitan, the cultivated Easterner from the Maritime

provinces, English in background, well educated, an elocution-
ist by training, a flamboyant figure in theater circles who toured
Canada giving dramatic readings. Despite her many absences, it
was she who ruled the family, and it was she who steered both
Marshall and his younger brother, Maurice, who became a
Presbyterian minister, toward intellectual careers. Since not even
star elocutionists, much less so-so insurance salesmen in
Western Canada, made a lot of money, the McLuhans lived
modestly, but Elsie McLuhan would make sure, in due course,
that her son Marshall, the academic star, was educated abroad.
In 1920, when he was nine, the family moved from Edmonton
to Winnipeg, and he went to high school and college there,
graduating from the University of Manitoba, which was about
a mile from his house, with a bachelor's degree in 1932 and a
master's degree in English literature in 1933. His mother,
however, had grander credentials in mind. At her prodding, he
applied for and won a scholarship to Cambridge University
in England.

At this point McLuhan was very much the traditional
young scholar, "the literary man," a type he would later
ridicule as smugly ignorant of the nature of the very medium
he studied and labored in, namely, print. As it turned out, in
the 1930s the literary life at Cambridge, at Oxford, and
in London was anything but traditional. This was the trough
of the Great Depression, and British intellectuals had begun to
take an interest in the lower orders, "the masses," many
as Marxists but others as students of what would later be
called popular culture. McLuhan was drawn to the work of
Wyndham Lewis and the Cambridge scholar F. R. Leavis, who
were treating movies, radio, advertisements, and even comic
strips as a new "language."

These were also the palmy days of Catholic writers such as
Hilaire Belloc and G. K. Chesterton, whose wit and sophistica-
tion had suddenly made Catholicism exciting, even smart, in

literary circles. Two of the most brilliant and seemingly cynical of the London literati, W. H. Auden and Evelyn Waugh, converted to Catholicism in this period. Likewise, Marshall McLuhan. He became a convert to the One Church – and to the study of popular culture. Although almost nothing in McLuhan's writing was to be overtly religious, these two passions eventually dovetailed to create McLuhanism.

After receiving a second bachelor's degree from Cambridge, he began his teaching career in 1936 in the United States, at the University of Wisconsin. He returned to Cambridge in 1939 and over the next three years received a master's degree and a doctorate in English literature. After Wisconsin, he taught only in Catholic institutions, first the University of St. Louis, then at Assumption University in Windsor, Ontario, and finally at the Catholic college of the University of Toronto, St. Michael's, where he joined the faculty in 1946.

By this time Marshall McLuhan was thirty-five years old and the very embodiment of Elsie McLuhan's appetite for things intellectual – and for the center of the stage. He was known both as a literary scholar, an expert in sixteenth- and seventeenth-century English literature and the work of James Joyce, and as a charismatic figure who captivated groups of students and faculty with his extracurricular Socratic gatherings devoted to "the folklore of industrial man," as he called it, in which he decoded what he saw as the hidden language of advertisements, comic strips, and the press. He would show a slide of a Bayer Aspirin ad featuring a drum majorette wearing a military helmet and jackboots and carrying a baton the size of a mace. The caption reads, "In 13.9 seconds a drum majorette can twirl a baton twenty-five times . . . but in only TWO SECONDS Bayer Aspirin is ready to work!" What is the true language of such an ad, he would ask? What does it really convey? Why, a "goose-stepping combination of military mechanism and jackbooted eroticism," the wedding of sex and technology, a recurring advertising theme he christened "the mechanical bride."

That was the title of his first book, published in 1951, when he was forty years old. *The Mechanical Bride* had the conventional anti-business bias of the literary man, aimed, as it was, at liberating the public from the manipulations of the advertising industry; but it also led McLuhan into the orbit of his colleague at Toronto, the economic historian Harold Innis. As McLuhan himself was quick to point out, it was from two books published by Innis in 1950 and 1951, *Empire and Communications* and *The Bias of Communication*, that he drew the central concept of McLuhanism: namely, that any great new medium of communication alters the entire outlook of the people who use it. Innis insisted that it was print, introduced in the fifteenth century by Johann Gutenberg, that had caused the spread of nationalism, as opposed to tribalism, over the next five hundred years. McLuhan published his first major theoretical work, *The Gutenberg Galaxy*, in 1962, when he was fifty-one. He called it "a footnote to the work of Harold Innis."

His master stroke came two years later when he brought the Innis approach forward into the twentieth century and the age of television with *Understanding Media*. McLuhan theorized that print had stepped up the visual sense of Western man at the expense of his other senses, which in turn led to many forms of specialization and fragmentation, from bureaucracy, the modern army, and nationalistic wars to schizophrenia, peptic ulcers, the cult of childhood, which he regarded as fragmentation by age, and pornography, the fragmentation of sex from love. In the second half of the twentieth century . . . enter television. Television, said McLuhan, reverses the process and returns man's five senses to their pre-print, pre-literate "tribal balance." The auditory and tactile senses come back into play, and man begins to use all his senses again in a unified "seamless web" of experience. Television, McLuhan maintained, is not a visual medium but "audio-tactile." This was the sort of contrary utterance he delighted in making, contradicting common sense without bothering to explain or debate. The world, he said, was fast becoming "a global village," that being the end result of television's seamless web spreading over the earth.

The immediate effects of television on the central nervous system, said McLuhan, may be seen among today's young, the first television generation. The so-called generation gap, as he diagnosed it, was not ideological but neurological, the disparity between a print-bred generation and its audio-tactile, neo-tribal offspring. McLuhan was observing the new generation up close. In the summer of 1939 he had been in California visiting his mother, who was teaching at the Pasadena Playhouse, when he met an American actress, Corinne Lewis, fell in love with her, proposed to her then and there, married her on the spot, and took her off to Cambridge, all in such a short order that she had to wire her parents to let them know she was now Mrs. McLuhan. Marshall and Corinne McLuhan had six children, four daughters and two sons. Personally, McLuhan had little patience with television or any other electronic medium, but he looked on with awe as his children seemed to study for school, watch television, talk on the telephone, listen to the radio, and play phonograph records all at the same time. The new generation, he was convinced, was bound to sit baffled and bored in classrooms run by print-bound teachers. This, he argued, meant the educational system must be totally changed.

But then the new sensory balance was going to bring about Total Change – he used a capital T and a capital C – in any case. Just as the wheel was an extension of the human foot, said McLuhan, and the axe was an extension of the arm, the electric media were extensions of the human central nervous system, and these nervous systems would be brought together in an irresistible way. His predictions were not tentative. Human nature would now be different. Nationalism, the product of print, would become impossible. Instead: the global village. In the global village, he predicted, it would no longer be possible to insulate racial groups from one another. Instead, all would be "irrevocably involved with and responsible for" one another. McLuhan warned that the global village was not a prescription for utopia. In fact, it might just as easily turn out to be a bloodbath. After all, he asks, where do we find the most accomplished butchers? In villages. The global village could bring all humanity together for slaughter as easily as anything else.

Yet he also believed the new age offered the possibility of something far more sublime than utopia, which is, after all, a secular concept. "The Christian concept of the mystical body," McLuhan wrote in one of the few explicit references to his fondest dream, "of all men as members of the body of Christ – this becomes technologically a fact under electronic conditions."

And here we see the shadow of the intriguing figure who influenced McLuhan every bit as much as Harold Innis but to whom he never referred: Pierre Teilhard de Chardin. Teilhard de Chardin was a French geologist and paleontologist who first made a name for himself through fossil-hunting expeditions in China and Central Asia. At the age of thirty, in 1911 (the year, it so happens, McLuhan was born), he became a Jesuit priest and taught geology at the Catholic Institute in Paris. His mission in life, as he saw it, was to take Darwin's theory of biological evolution, which had so severely shaken Christian belief, and show that it was merely the first step in God's grander design for the evolution of man. God was directing, in this very moment, the twentieth century, the evolution of man into a noösphere – that was Teilhard de Chardin's coinage, a noösphere – a unification of all human nervous systems, all human souls, through technology. Teilhard (pronounced Tay-yar, as he was usually referred to) mentions radio, television, and computers specifically and in considerable detail and talks about cybernetics. Regardless of what anybody thought of his theology, the man's powers of prediction were astonishing. He died in 1955, when television had only recently come into widespread use and the microchip had not even been invented. Computers were huge machines, big as a suburban living room, that were not yet in assembly-line production. But he was already writing about "the extraordinary network of radio and television communication which already link us all in a sort of 'etherised' human consciousness" and of "those astonishing electronic computers which enhance the 'speed of thought' and pave the way for a revolution in the sphere of research." This technology was creating a "nervous system for humanity," he wrote, "a single, organized, unbroken membrane over the earth," a "stupendous thinking machine." "The *age of civilization* has ended, and that of <u>one civilization</u>" – he underlined <u>one civilization</u> – "is beginning." That unbroken membrane, that noösphere, was, of course, McLuhan's "seamless web of experience." And that "one civilization" was his "global village."

We may think, wrote Teilhard, that these technologies are "artificial" and completely "external to our bodies," but in fact they are part of the "natural, profound" evolution of our nervous systems. "We may think we are only amusing ourselves" by using them, "or only developing our commerce or only spreading ideas. In reality we are quite simply continuing on a higher plane, by other means, the uninterrupted work of biological evolution." Or to put it another way: "The medium is the message."

Privately McLuhan acknowledged his tremendous debt to Teilhard de Chardin. Publicly he never did. Why? For fear it would undercut his own reputation for originality? That would

have been very much out of character. After all, he acknowledged his debt to Harold Innis openly and on his knees in
gratitude. The more likely reason is that within Catholic
intellectual circles – and we must remember that McLuhan
was on the faculty of the University of Toronto's Catholic
college, St. Michael's – Teilhard de Chardin was under a
cloud of heterodoxy. Decades earlier the Church had forbade
him from teaching or publishing his theory of evolution, since
he accepted most of Darwinism as truth. None of his six
books on the subject was published in his lifetime. But among
intellectuals at St. Mike's, as they called St. Michael's College,
there was a lively underground, a Jesuit *samizdat* in Teilhard
de Chardin manuscripts, especially after he moved to the
United States in 1951. McLuhan was fascinated by Teilhard
but he presented a problem. Even in death he remained out of
the bounds of Catholic theology, and McLuhan took his faith
very seriously, all the more so because he was a convert from
Protestantism teaching in a major Catholic institution.

But Teilhard presented a secular problem as well. McLuhan
was living in an age in which academic work with even a tinge
of religion was not going to be taken seriously. Inside the
Church, Teilhard may have been considered too much of a
Darwinian scientist, but outside the Church he was considered
too much of a Catholic mystic. When *Understanding Media*
was published in 1964, it was loaded with Teilhard de
Chardin, but it would have taken another Teilhard enthusiast
to detect it, and a subtle one at that. Not a single theological
note was struck.

Indeed, *Understanding Media* exploded upon the intellectual world in the mid-1960s with a distinctly earthly brilliance
and immediately caught the attention of many of the most
devoutly materialistic and practical minds in commerce
and industry. In part it was the deceptively simple title,

Understanding Media, which came across as a challenge: "You people who use the media, who own the media, who invest millions in the media and depend on the media – you don't begin to understand the media and how they actually affect human beings." By late 1964, corporations such as General Electric and IBM were inviting McLuhan to the United States to talk to their executives. Their attitude was not so much "He's right!" as "What if he is right? (We'd better find out.)" McLuhan informed General Electric that they might think they were in the business of making light bulbs, but in fact they were in the business of moving information, every bit as much as AT&T. Electric light was pure information, a medium without a message. IBM he somewhat condescendingly praised for having finally realized that they were not in the business of manufacturing equipment but of processing information. He excelled at telling powerful and supposedly knowledgeable people they didn't have the foggiest comprehension of their own enterprises. He never adopted a tone of intentional shock, however. He was always the scholar, speaking with utter seriousness. He had a way of pulling his chin down into his neck and looking down the nose of his long Scottish-lairdly face before he delivered his most delphic pronouncements. He seemed to exist out beyond and above them all, surveying them from a seer's cosmic plane.

But what turned Marshall McLuhan from a University of Toronto English professor with an interesting theory into McLuhan, a name known worldwide, was the curious intervention of a San Francisco advertising man, Howard Gossage. Fascinated by *Understanding Media*, Gossage took it upon himself, at his own expense, to become McLuhan's herald, bringing him to the United States in 1965 and introducing him to the press and the advertising industry on the West Coast and in New York. It proved to be a brilliant campaign. Magazine articles, newspaper stories, and television appearances were generated at an astonishing rate. Late in 1965 both *Harper's Magazine* and *New York Magazine* published major pieces about McLuhan. In the single year 1966 the number grew to more than 120, in just about every important publication in the United States, Canada, and Great Britain. The excitement was over the possibility that here might be a man with an insight of Darwinian or Freudian proportions.

As his fame grew, so did the ranks of his detractors, particularly among literary people, whom he regularly wrote off as hidebound, reactionary, and oblivious of how even their own medium, print, actually worked. Scientists, meantime, didn't know what to make of him one way or the other. The heart of his theory, the concept of the human "sensory balance," falls within the field of cognitive psychology or, more broadly, neuroscience. Today neuroscience is the hottest subject in the academic world, but even now there is no way of determining whether or not any such balance exists or whether or not a medium such as television can alter one individual's nervous system, let alone an entire society and the course of history. McLuhan treated any and all critics with a maddening aloofness. He was not trying to create a self-contained body of theory, he insisted – although in fact he probably was – he was a pioneer heading out into a vast *terra incognita*. So little was known, and there was so little time. His mission was to explore, to make the "probes," to use one of his favorite words, to open up the territory. Others, those who came after, could conduct the systematic investigations, run the clinical experiments, organize the data, and settle the disputes. He dismissed all opposition as what Freud called "resistance," a reluctance to let go of the comfortable notions of the past in the face of brilliant new revelations about the nature of the human animal.

In the wake of all the excitement over *Understanding Media* McLuhan established the Centre for Culture and Technology at the University of Toronto. This was an imposing laboratory-like name for what was, in fact, little more than a letterhead, a desk, the lined paper on which he wrote, by hand, and his amazingly fertile and facile mind. In this respect McLuhan was like Sigmund Freud. Very little of what Freud had to say has survived the scientific scrutiny of the past half-century. In hindsight we can see that he was a brilliant philosopher of the old school who happened to live in an age in which only science was accepted as gospel truth. So by night he led his philosophical speculations in through the back door of his clinic, and in the morning he marched them out the front door as scientific findings. Thus also McLuhan at the Centre for Culture and Technology. At bottom, McLuhan remained, through it all, a literary man in the grand tradition of Samuel Johnson, Thomas Carlyle, Matthew Arnold, and G. K. Chesterton, with the gift of brilliant flashes of insight into the era in which he lived.

He never endeared himself to literary people, however, because so many of his wittiest, Chesterton-like sayings were at their expense. Asked to comment on the headlong rush of writers and scholars into protest movements during the 1960s, he said: "Moral bitterness is a basic technique for endowing the idiot with dignity."

In the mid- and late-1970s, the mocked had their revenge. McLuhan didn't seem to realize that an academic celebrity, if he wants to maintain his worldly eminence, is compelled to act oblivious of, or at least utterly aloof from, the journalists, show biz folks, and publishers who so merrily magnify his reputation to star status. Freud and Einstein understood this very well. In 1922 the *Chicago Tribune* offered Freud $25,000, the equivalent of $300,000 today, to come to the United States and provide psychoanalytical commentary the *Tribune* could run during the trial of the "thrill-killers" Leopold and Loeb. The bearded one wasn't about to. He came to the United States only to give an abstruse lecture at City Desk–proof little Clark University in Massachusetts. McLuhan, in contrast, published co-written books with jokey titles such as *The Medium is the Massage* and let Woody Allen put him in the movie comedy *Annie Hall* playing himself, in cameo, as a pun-cracking, recondite theorist. By the time he died at the age of 69 in 1980 after a series of strokes, his critics, chiefly New York intellectuals, had successfully nailed him as "not serious" and therefore over and done with.

Yet McLuhan had introduced a notion that the *fin de siècle*'s fast-proliferating breed of young computer techies would not let die, namely, the idea that new media such as television have the power to alter the human mind and thereby history itself. 1992 came – bango! – a new medium, computers linked up to telephone lines to create an Internet. The Internet lit McLuhanism up all over again, and the man himself was resurrected as something close to a patron saint. He was certainly that to the edgiest and most prominent of the new dot-com journals, *Wired*, which ran his picture near the masthead in every issue.

Dear God – if only Marshall had been alive during the 1990s! What heaven those ten years would have been for him! How he would have loved the Web! What a shimmering Oz he would have turned his global village into! Behold! The fulfillment of prophecies made thirty years before! The dream of the mystical unity of all mankind – made real!

Of course, no sooner had the third millennium begun than the dot-com bubble burst and McLuhan's young Silicon Valley apostles awoke with a shock. They shook their heads to clear them and tried to refocus their vision of the future. Many could not. But a Gideon's army of the young could make out a tiny Halogen bulb, no bigger than a traveling-size toothpaste cap, still burning . . . and its light shone 'round about them . . . and they say it still does.

New communications theorists will arise, as if from straight out of the asphalt, the concrete, the vinyl tiles, or the PermaPour flooring. But one thing will not change. First they will have to contend with McLuhan.

Preface

by Stephanie McLuhan

This book is the brainchild of David Staines, who recognized the literary potential of the audiovisual material from which it is derived. Although as co-editors we approached the project from different perspectives – David is a literary historian while I am a television producer – we have in common an intimate knowledge of the subject. As well, our longstanding friendship made the collaboration remarkably comfortable.

There is no first time I can remember meeting David. He visited our home many times over the years. It seemed logical, therefore, to call him for advice about what to do with the roughly two dozen tapes of Marshall McLuhan's lectures and television interviews I had collected over twenty-five years. I hoped that David would recommend the ideal university communications department where I could donate them; instead, he enthusiastically said they would make an important book. I was surprised by his suggestion and insisted that we get together to screen the tapes before we made up our minds. When we had done this, he summarized the proposed book: it would consist only of primary-source material that had never been published before, and be based on unedited lectures and television interviews electronically recorded over twenty years from 1959 until 1979. It seemed straightforward, and so we agreed to commit ourselves to the venture.

After many months, much brainstorming, and a few wrong turns, we finished with our relationship intact. There were some exasperating moments that usually occurred when we were poring over transcripts of the various lectures, which were considerably more difficult to absorb than the television interviews. It would take three or even four readings of a piece to comprehend it fully, mainly because it was packed with so many thoughts and ideas.

In the eighteen selections, which are presented chronologically, there are, of course, ideas that come up a number of times, but it is engaging to follow the development of McLuhan's thought process through the years. His views on his own work and on the world are valuable adjuncts to his publications.

Taken together, these lectures and interviews make up a biography/autobiography enabling you to read Marshall McLuhan in the original where you will find a more accessible, even unmediated encounter than is possible through his books.

In the footnotes we have attempted to identify every quotation. In a few cases, however, we could not locate precise information.

Our sincere appreciation is extended to Tom Wolfe for his masterful Foreword, to the broadcasters and interviewers who created the interviews, and to the university archivists who provided background information on the lectures.

Electronic Revolution:
Revolutionary Effects of New Media
(1959)

On March 3, 1959, Marshall McLuhan addressed a gathering of more than a thousand educators in Chicago sponsored by the American Association for Higher Education. The theme of the conference was "The Race Against Time: New Perspectives and Imperatives in Higher Education," and McLuhan's talk was titled "Electronic Revolution: Revolutionary Effects of New Media."

The forty-seven-year-old McLuhan had already published The Mechanical Bride: Folklore of Industrial Man (1951), his shrewd dissection of the manipulative techniques of the advertising industry. By 1959 he had become known in academic circles and beyond as a pioneering thinker on the mass media.

In this address he speaks as an educator to an audience of educators: "So rapidly have we begun to feel the effects of the electronic revolution . . . that all of us today are displaced persons living in a world that has little to do with the one in which we grew up." The electronic revolution of television has made the teacher the provider no longer of information but of insight, and the student not the consumer but the co-teacher, since he has already amassed so much information outside the classroom.

Today in the post-mechanical age, we are in the same position as horse-minded people when confronted with the automobile. To horse-minded people the most striking fact about the car is that it is a horseless carriage. In the same way, radio appeared as wireless to those who had become accustomed to the miracle of the telegraph. Automation to machine-minded people strikes fear as being an extreme form of mechanization; but as Peter F. Drucker says in his *Landmarks of Tomorrow*, automation "is merely a particularly ugly word to describe a new view of the process of physical production as a configuration and true entity."[1]

So rapidly have we begun to feel the effects of the electronic revolution in presenting us with new configurations that all of us today are displaced persons living in a world that has little to do with the one in which we grew up. Most of us can recall the days when children pushed hoops along sidewalks and roads. There are more hoops than ever now. But no child will push one. For children today live in a space whose configurations are not those of thirty years ago. Instead of being attracted by an outer space designed in lineal fashion, children now nucleate their own space, ballet-style. Living, for example, with electronic imagery in which the image is formed by light *through* rather than light *on* (one major difference between TV and film), children respond with new sensory configurations and new attitudes to their world.

Educators naturally feel that their job is to maintain the educational establishment, and to preserve and advance the values so long associated with its procedures. Right now this means, for example, that we are going to insist that Johnny acquire the art of reading, if only because print is the matrix of

Western industrial method in production, and print teaches consumer habits and outlook as well. Print teaches the habit of sequential analysis and of fragmentation of all motion into static units. Print teaches habits of privacy and self-reliance and initiative. It provides a massive visual panorama of the resources of our mother tongue which preliterate peoples know only by ear. In fact, print is not only access to our culture and technology, it is our culture and technology. That is why in the electronic age we are threatened by new fast-moving and flexible media – while we sit in a Maginot Line convinced of the importance of our position.

Of course Johnny must read. He must follow the lines of print. He must roll that hoop down the walk. He must roll his eyes in lineal, sequential fashion. We have only to proceed to engraft the old right-handedness on his new left-handedness in order to win our point. But in the meantime we shall have lost his attention, and he may be subdued, but he will be utterly confused.

Taken in the long run, the medium is the message. So that when, by group action, a society evolves a new medium like print or telegraph or photo or radio, it has earned the right to express a new message. And when we tell the young that this new message is a threat to the old message or medium, we are telling them that all we are striving to do in our united social and technical lives is destructive of all that they hold dear. The young can only conclude that we are not serious. And this is the meaning of their decline of attention.

I have said that the medium is the message in the long run. It would be easy to explain and confirm this point historically. Print simply wiped out the main modes of oral education that had been devised in the Greco-Roman world and transmitted with the phonetic alphabet and the manuscript throughout the medieval period. And it ended that 2,500-year pattern in a few decades. Today the monarchy of print has ended, and an oligarchy of new media has usurped most of the power of that five-hundred-year-old monarchy. Each member of that oligarchy possesses as much power and message as print itself. I think that if we are to have a constitutional order and balance among these new oligarchs, we shall have to study their configurations, their psychodynamics, and their long-term messages. To treat them as humble servants (audiovisual aids) of our established conventions would be as fatal as to use an X-ray unit as a space heater. The Western world has made this kind of mistake before. But now with the collapse of the East, that is, with its recognition that no viable society can be built anywhere except on Western modes, it would be a very bad time to allow our own new media to liquidate the older media. The message and form of electronic information pattern is the simultaneous. What is indicated for our time, then, is not succession of media and educational procedures, like a series of boxing champions, but coexistence based on awareness of the inherent powers and messages of each of these unique configurations.

In his book on *Film as Art* Rudolph Arnheim, the psychologist, wrote: "The history of human ingenuity shows that almost every innovation goes through a preliminary phase in which the solution is obtained by the old method, modified or amplified by some new feature."[2]

In the past thirty years all of our traditional disciplines in the arts and sciences have moved from the pattern of lineal cause to configuration. Nowhere is this more true than in biology. Yet the methods used to reach configuration are still the old Cartesian methods of classical mechanics applied to the study of living organism. And configuration concepts such as stress or metabolism ecology and syndrome are essentially aesthetic terms.

As we move into the world of the simultaneous out of the era of mechanism and of the lineal succession types of analysis, we not only move into the world of the artist, but we see the disappearance of the old oppositions between art and nature, business and culture, school and society. It really does not matter to which phase of our culture today we turn. The habit of simultaneous vision of all phases of process is what characterizes the articulate awareness in the field.

Thus, in the movement of information today by technological means we have by far the largest industry. American Telephone and Telegraph alone greatly exceeds the capitalization of General Motors. The production and consumption of information, that is, is the main business of our time. Culture has taken over commerce. Within industry itself the growth of the classroom for workers and for management receives a budget at least three times the $16 billion budget of formal education in North America. And for research also, the trend and ratios are similar.

The movement of information round-the-clock and round-the-globe is now a matter of instantaneous configuration. Decision-making in business and in education as much as in diplomacy is now a matter of grasping these configurations. They have a language and a syntax of their own as much as does the iconology of pictorial advertisement, so that it is not only the business of education today to teach these new languages, but to teach how we can in our previously achieved configurations of culture be enriched by these new powers and not merely dissolved by them. There is a classic definition of science originating in the Académie française after the death of Descartes: "The certain and evident knowledge of things by their causes." Survival indicates that we grasp by anticipation the inherent causes and not the effects of the electronic media in all their cultural configurations and make a fully conscious choice of strategy in education accordingly.

The eminent French anthropologist, Claude Lévi-Strauss, in an analysis of "The Structural Study of Myth," presents us with the typical configurational insight: "We define myth as consisting of all its versions . . . therefore, not only Sophocles, but Freud himself, should be included among the recorded versions of the Oedipus myth on a par with earlier or seemingly more 'authentic' versions."[3] Applied to the study of media in education, the Lévi-Strauss insight, which is characteristic of the approaches of the arts and sciences in our time, means that we have to regard our media as mythic structures, as massive codifications of group experience and social realities. And just as print profoundly altered the structure of the phonetic alphabet and repatterned the educational processes of the Western world, so did the telegraph reshape print as did the movie and radio and television. These structural changes in media myth coexist in an ever-live model of the learning and teaching process. The changing configurations of this massive structure inevitably alter the bias of sight, sound, and sense in each one of us, predisposing us now to one pattern of preference, and now to another. Today, via electronic means, the coexistence of cultures and of all phases of process in media development offers to mankind, for the first time, a means of liberation from the sensory enslavement of particular media in specialized phases of their development.

What Harold Innis well called *The Bias of Communication* concerned not only the forms in which men have chosen to codify information but also the causal effects of stone, papyrus, and print on the changing structures of decision-making.

Mr. Parkinson has recently entertained us with an analysis of bureaucratic decision-making as it exists in the written mode of the memorandum syndrome. The written forms of information movement begin to look quaint after a few decades of electronic information pattern. At present the co-pilots of Canadian jet fighters have to make decisions in quite another configuration, namely, that of the instantaneous. Before being assigned to their common task, they undergo a long phase of what is called "going steady." When finally assigned to their plane they are publicly "married" by the commanding officer in a sober ceremony. Today, it is felt only marriage can connote the degree of togetherness, tolerance, and sympathy necessary for decision-making in the use of new technology. This new pattern is the subliminal but overwhelming message of the media since the telegraph. Yet nowhere in our educational establishment have we made provision for the study of these profound messages which impose their configurations on the sensory equipment of children from their first days of existence. Yet some such provision would seem to be indicated against the persistent effects of media fallout.

One effect of the commercial movement of information in many media is that today we live in classrooms without walls. The printed book created the classroom as we know it by making available exactly repeatable information. Even if the manuscript or handmade book had been cheap enough for all, it could never have been uniform or repeatable. Moreover, the best manuscripts are slow to read and create a totally different feeling for language in the student – a feeling for the multiple layers of meaning. Such a feeling has returned today, especially since television, with its light *through* rather than light *on* the image. In a word, the printed page was no more a cheaper manuscript than the motorcar was a horseless carriage. And the repeatable character of print had consequences in science and industry which we are still working out.

But all previous configurations, including that of print from movable type, undergo a sort of alchemical change when they meet a heavy new stress or pull from a new type of configuration.

I have called the electronic age, which began with the telegraph, the post-mechanical age. For now, that which moves in our new structures is no longer wheels and shafts (except incidentally), but light itself. We can now see in depth the shape of the Gutenberg myth and technology. Our knowledge of the causal operation of the Gutenberg configuration might now save the Indians and the Chinese a great deal of needless liquidation of many elements of their cultures, which we have come to value in the West. But even more urgently we need prescience of the full causal powers latent in our new media in order that we may do for our own print culture what we could also do to save Chinese ideogrammic calligraphy and education. A kind of alchemical foreknowledge of all the future effects of any new medium is possible. Under electronic conditions, when all effects are accelerated in their mutual collision and emergence, such anticipation of consequence is basic need as well as new possibility. For example, our present concern about closed-circuit television in education is parallel to the sixteenth-century concern about whether print and the vernaculars could do a serious educational job. It is actually asking whether the car can ever supplant the horse. We are losing precious time in such static retrospection.

Let me mention one central feature of the electronic configuration, namely, its strong tendency to reverse producer-consumer relationships. Print over the centuries had stabilized a pattern of producer-consumer relations. But with the telegraph a century ago the reader of the press had to assume an editorial function unknown to the reader of the pre-telegraph press.

When news moves slowly, the paper has time to provide perspectives, background, and interrelations for the news, and the reader is given a consumer package. When the news comes at high speed, there is no possibility of such literary processing, and the reader is given a do-it-yourself kit. This telegraph pattern was soon transferred to poetry, painting, and music, to the bewilderment of consumer-oriented people. When John

Dewey attempted to transfer the same electronic or do-it-yourself pattern to in-school education, he failed. He had not analyzed the situation adequately nor had he any glimpse of the media factors operative on his own enterprise. But had he merely turned the do-it-yourself bias towards the training of the young in the perception and judgment of the out-of-school media, he would have succeeded, and we would all of us be in a much stronger position educationally today because that is precisely the task we must now tackle – the training of the young in mastery of the new global media.

Most of my remarks so far have been pointing out the mere nature of the technological causes which, past and present, produce change in educational patterns. These causes are mainly subliminal and non-verbal. And may it not be that the new importance that is now accorded to the arts, both in education and in industry, is owing to our awakened sense of the role of art and artists in raising subliminal and non-verbal factors of experience to the level of conscious articulation?

In a simultaneous information structure such as the electronic global community, we cannot afford subliminal factors since their operation is haphazard. The simultaneous compels us to make a social order that, like a poem or painting, is totally realized in its interrelations, and in which each factor has total relevance.

To record briefly some basic educational changes which are now discernible and may well foreshadow major lines of development, let me suggest the following:

We have, in the age of literacy, educated more and more members of society. In the electronic age we shall educate more of each person. We now move from educational extension to even greater extension, but in depth as well.

Is not this the drift of our new concern with the gifted child?

The meaning of the New Criticism today is not just literacy but a shift to reading in depth with total awareness rather than the single-plane approach of the older literacy.

As we extend our educational operation by television and videotape we shall find that the teacher is no longer the source of data but of insight. More and more teachers will be needed for the type of depth instruction that goes naturally with television, with light *through* rather than light *on*.

The need for more and more profound teachers because of the very medium of television is shadowed in the panel show, at least to the extent that it seems more natural, even since radio, than a single source of comment and information. Two or more teachers in dialogue with each other and with class or audience create exactly that sense of light *through* rather than light *on*, which is the nature of the television image or mosaic as compared with movie or print. In the same way with the panel, the voice comes through the audience rather than to the audience.

In the same way that industry now makes the consumer the producer by means of motivation research, do not educators now recognize the education problems to be motivation rather than consumption of packaged information? The fully motivated student is creative in his consumption and cognition. He is co-author and co-producer, so that the new teaching must increasingly cast the student in co-teacher roles. And, indeed, he is already potentially in such a position because of his vast intake of information in out-of-classroom experience, which is only in part shared by the teacher.

Increasingly the business of education will be discovery and interrelation. And just as industrial production now depends entirely on higher education, and as culture has become the main business of the globe, so learning and not teaching may well become the most highly paid profession. As we begin to learn for participation, rather than for specialist, applied knowledge patterns of action, we can look back and see how the growing habit of conferences already forecasts this change in the roles of teacher and learner. Applied knowledge for production is now taken for granted and knowledge shifts to the global role of community and participation in a way commensurate with the roles of the new media.

1. Peter F. Drucker, *Landmarks of Tomorrow* (New York: Harper & Brothers, 1959), p. 5.
2. Rudolph Arnheim, *Film As Art* (London: Faber & Faber, 1958), p. 146.
3. Claude Lévi-Strauss, "The Structural Study of Myth," *Journal of American Folklore* 68 (1955), p. 435.

Popular/Mass Culture:
American Perspectives
(1960)

In 1959–60, on sabbatical leave from the University of Toronto, McLuhan served as director of the Project in Understanding New Media for the National Association of Educational Broadcasters in Washington, D.C. He was an extraordinarily prolific writer of scholarly articles, and by this time his views on the electric media were already so well-known in academic circles, he was the central figure at the third annual Conference on the Humanities on October 28 and 29, 1960, sponsored by the Ohio State University's Graduate School. The general subject of the meeting was "Popular/Mass Culture: American Perspectives."

On the first day, McLuhan gave a lecture titled "Technology, the Media, and Culture." At this stage in his career, McLuhan was unabashedly optimistic about the potential of the new electric media: "The emergence of a global community of learning is a natural outcome of a world in which the production and trans-portation of commodities finally merges with the movement of information itself."

On the second day, McLuhan participated in a panel discussion chaired by Gilbert Seldes (1893–1970), the leading cultural critic of the day. The theme was "The Communications Revolution." Joining McLuhan on the panel were two professors of communica-tions at Ohio State University: Edgar Dale (1900–1985), professor

of audiovisual education, and Keith Tyler (1905–94), professor of
radio education. During their conversation McLuhan articulates
one of his most famous theorems, namely, that television is a cool
medium that will not tolerate hot characters.

■ ■ ■ ■ ■

TECHNOLOGY, THE MEDIA, AND CULTURE

A member of Harvard's faculty of Far Eastern studies was
dining some years ago with a mandarin friend in Peking. He
was both pleased and puzzled to note that the room was
adorned with American pin-ups, with college pennants, Coke
bottles, book-matches, and advertisements from American
magazines. The Harvard man spoke with enthusiasm about the
mandarin's broad interest in American life as here displayed.
But the mandarin said he merely wished to reciprocate the
courtesy shown to him in America by his many friends whose
walls and mantles were filled with the coolie art of prints and
domestic trash from Eastern bazaars.

In the same way, we may have been baffled to hear that
Picasso has always had a keen regard for American comic-strip
art and James Joyce surrounded himself with materials of the
most popular songs and journalism. Flaubert was Joyce's
master in the poetic scrutiny of the stereotypes of the arts of
mass appeal. His analysis of these new popular forms in the
middle of the last century justified his saying that if people had
read and understood his *Sentimental Education* there would
have been no war of 1870. In the same way, Wyndham Lewis
observes that if people had understood his analysis of popular

culture in *The Art of Being Ruled*, there would have been no World War II.

What these men were saying about their serious scrutiny of the popular arts is now being said by John Kenneth Galbraith in the current issue of *Horizon* [Sept. 1960]. Writing about "The Muse and the Economy," Galbraith makes in relation to business the point that for more than a century has been palpable to the artists of the Western world, namely, that for the decision-maker, popular taste affords no timely data. But the experimental artist is all the time building models of future situations which afford reliable beacons for the social navigator. The social scientist can only report on current patterns of taste, he has no access to future patterns such as the artist has always had. And the reason for this is simply that the artist, as Wyndham Lewis said to me, "is engaged in writing a detailed history of the future because he is aware of the unused potential of the present."

The very next article in the same issue of *Horizon* is by Russell Lynes, who in *Life* magazine a few years ago reported a reversal of all known laws of nature and economics in the emergence of the penniless intellectuals of America as holding the whip-hand in the control of design of consumer goods. Our very topic for this conference illustrates a hardening of the categories of thought and perception in terms of consumer goods. Nothing is more characteristic of the highly literate than the assumption that the difference between popular culture and elite culture is a difference in the type of product that is consumed. He takes it for granted that "by their cornflakes ye shall know them," not "by their fruits." The whole of our humanities programs have been structured on the consumer assumption of value. Our writers and poets and artists have had to learn producer orientation and creativity elsewhere, in the newspaper office, on Madison Avenue, and in Hollywood. And Europe has always been generous in recognizing the unique power and value of our artistic production in the popular commercial areas. They had had nothing to match it and are only just now entering the world of consumer values which we in North America are ready to abandon.

The book was the first mass-produced commodity. Print, by definition uniform and repeatable, not only created the very concept of "commodity" but made possible markets for such uniform and repeatable commodities. That the operation of the forms and matrices of the print assembly line when extended to all forms of production should also have shaped our attitudes to elite activities is quite natural. In England and America alike the elites have been lotus-eaters, elegant consumers of imported goods simply because a print-oriented world is a consumer world. During the past century our mechanical print galaxy has been moving into an electric galaxy with a resulting reconfiguration of patterns even among the familiar components. The electric galaxy is producer-oriented, rewiring our cultural circuit, and throws malign lustre on our traditional consumer values. This process which began with telegraph has reached full proportions with television. Our teenagers assuming the artist outlook have rejected the consumer world. And even in business, as so popular a writer as John Kenneth Galbraith has testified, the old quarrel between art and commerce has ended in a wedding. We are now ready for a peripeteia in the Western drama. Having long admired the spontaneity and art of backward men in preliterate and semi-literate societies, we now find ourselves well on the road to retribalization via our new electric media. Having long talked of the plight of the individual in a mass society we can now get ready to write about the plight of mass man in an individualist world. Even the wheel, the basis of Western mechanical enterprise, may in the jet age reamalgamerge with the animal form from which it originally was abstracted.

The entire drama of conflict between individual and mass is most usefully studied under the aspect of the role of a poet in relation to his medium, because a language is a mass medium in all senses. Nobody in particular made it. Yet individuals have always to think and dream and feel in terms of this mass medium. The poet is in a special way the custodian and rejuvenator of language.

Wyndham Lewis, the painter and writer, devoted much of his energy to the study and delineation of the Western drift back into the "sacred" auditory space of primitive and irrational man. He repudiated the Spenglerian picture of this development. In place of Spengler's popular notion of inner cosmic necessity, Lewis placed the responsibility for the trek from rational, visual values squarely at the door of artists and scientists and philosophers who were climbing aboard the bandwagon of popular mass media. That is to say, Lewis diagnosed the fondness of the avant-garde painters and poets for newspaper and cinematic techniques as an intellectual failure and also as the abrogation of all moral responsibility to Western values. The Lewis critique of Joyce and Pound, for example, does not question their high artistic talent, but it derides their readiness to go along with the popular arts and trends of this century.

In the course of his indictment of our age as willing to abandon the entire heritage of the Greco-Roman achievement, Lewis brought most of the aspects and activities of the twentieth century under scrutiny. His work offers what is perhaps a more complete guide to the arts and letters of his age than that provided by any other writer in the history of literature. Recollecting in tranquility the many volumes of his work, it might be well for us to ask: "What avails the highest and most rigorous intellectual analysis directed to the very problems we have chosen to consider at this conference?"

When, in 1896, Bernard Berenson wrote, "The painter can accomplish his task only by giving tactile values to retinal impressions,"[1] he was not only very much aboard the impressionist bandwagon, he was advocating the television image. Unlike the movie image, the mosaic mesh of the TV bombards the viewer with tactile values. Popular technology would thus seem to be responsive to the highest behests of art. It was precisely in this type of matter that Lewis attacked his fellow artists for merely going along with technology. Lewis had the utmost contempt for the *Zeitgeist* and for those artists who try to discern the grimace of the *Zeitgeist* so that they can keep in line with it. He was not misled by any idea that the effects of new art forms could be mitigated by a flourish of noble "program content." Artists have always known that any art form has the power of imposing its own assumptions on the beholder. Any medium of communication is, like an art form, an extension of one or more of our senses. Speech alone is an extension of all of our senses at once. The mix or proportion of our senses made external to us ("uttered" = outered), the ratio or mix or proportion of our senses involved in speech or radio or photography, imposes non-verbally the parameters or frame of all human operations. The unspoken and even subliminal assumptions in any pattern of human association are dictated by the available means of codifying experience and of moving information. General awareness of this quite drastic fact seems to have departed from literate communities soon after the advent of the phonetic alphabet. When the alphabet was a revolutionary novelty, the Cadmus myth was formulated to explain the social operation of the alphabet. To wit, King Cadmus, who had introduced the Phoenician letters to Greece, had sowed the dragon's teeth, from which sprang up armed men. Myth would seem to be quite simply the perception and statement of a complex action of causes and effects in a single glance or gestalt. In the electronic age, when time and space factors are very much reduced in information flow, it is once more natural for us, as for men in small oral communities, to think mythically. For today it is easy to perceive consequences embedded in any kind of innovation. If we don't see them, they clobber us very quickly. In industrial design today, for example, the gap between product and consumer reaction has been reduced so far that they say with James Joyce: "His producers are they not his consumers?"[2] And instead of old-fashioned concepts of making the public aware of new products, they now speak of making the product conscious of its public, or target. The consequences of new media are perceived so fast that the dullest minds have begun to anticipate such effects by examining the forms or causes that are to be released on a public.

The largest item of industrial budgeting has for some time been the allocation for research. This is from necessity. An industrial galaxy is propelled so swiftly that it is constantly invading other galaxies with resulting stress and change of configuration. Peter F. Drucker points out in his book *Landmarks of Tomorrow* that it is no longer feasible in decision-making to exercise delegated authority, but only the authority of knowledge. When information moved slowly in written form, job specialism and pyramidal hierarchies of function were normal and even workable. The telephone and related electrical instruments have rendered the familiar organization patterns as obsolete as the assembly line. The latter has been liquidated by electric tape-recorded information flow, which coordinates with precision not one but whole clusters of operations. Richard Meier, in a paper given at Ann Arbor this past April ["Information, Resource Use, and Economic Growth"], formulated a natural law for media when he pointed out that increased levels of information flow result in substitutability:

With the elaboration of electrical engineering, and the fusing of many strands of chemical knowledge, a field that was evolving rapidly in a mainstream of its own that led from mass reactions to molecular, to atomic, and most recently to nuclear reactions, the possibility of a flexible, quick-acting, autonomous economy emerged. It is capable of substituting one set of raw materials by others so as to meet virtually all foreseeable emergencies which reduce or cut off supplies. . . . The task that remains is one of redesigning social institutions so that they are consonant with the revealed potentials of resource availability and technological efficiency.

With a parallel increase of accessibility of all cultures to all cultures and of all subject matters to all others, the redesigning of the educational establishments of the Western world is equally urgent not as an ideal but as a necessity. The older patterns of corporation management have had to be redesigned in the past ten years. And the new pattern is unmistakably nuclear or fieldlike as opposed to the old hierarchies of jurisdiction of staff and line and pyramidal functions.

The new pattern is one of small teams comprising clusters of diverse competencies with personnel accustomed to the crossing of functional lines in a perpetual dialogue of interpenetrating awarenesses. We have begun to see the emergence of such teams in the humanities divisions of our universities. But we have tended to assume that the overall structure of specialisms of our universities are still relevant to the tasks of teaching and learning. Soon we shall be engaged in historical consideration of just why current partitions and divisions of knowledge came to be established. Much in the same way, modern mathematics and physics have had to detach themselves from the assumptions and parameters of Euclidean space.

A book such as *The Sacred and the Profane* by Mircea Eliade is devoted to illustrating how both space and time are non-homogeneous and non-continuous to archaic man. That is to say, tribal man everywhere and at any time assumes the unique structuring of all spaces and times he encounters from moment to moment. Such an outlook is normal to painters since Cézanne and to poets since Baudelaire just as much as to the nuclear physicist. Today the problem is to explain that anomaly – Euclidean space, and its correlative, continuous time. Since no preliterate society ever had any experience of Euclidean space it is not too daring, I hope, to suggest that the fictions of Euclidean space may in a very special way owe their very existence to our Western experience with the phonetic alphabet. Hieroglyphic, pictographic, and ideogrammic modes of writing do not tend to bring into existence the abstract fictions of flat, straight, and uniform space. But the phonetic alphabet is an abstract technology for translating the multi-sensuous modes of speech into the merely visual. Letters are the language of civilization because they translate tribal man from his complex auditory and tactile world into a simple visual one which we have called "rational" ever since it was invented. On the other hand, number, the language of science, has been the means of translating the merely visual back into the sense of touch and sound. In his book *Number: The Language of Science*, Tobias Dantzig tells us that "The attempt to apply rational arithmetic to a problem in geometry resulted in the first crisis in the history of mathematics."[3] Today, that crisis is occurring on a massive cultural scale. Our rational Euclidean world of continuous and homogeneous space, extrapolated by the phonetic alphabet from the resonating tribal world, has now to face the electronic challenge of its own irrelevance and superfluousness. I think Dantzig can help us some more to get our bearings here. Just before the passage already quoted he is explaining the crucial use made in mathematics of the Renaissance concept of the "infinite process." If this concept does not derive from the new perception of perspective or vanishing point, it is at least parallel to it. "The prototype of all infinite processes," says Dantzig, "is *repetition*."[4] And this is a facet of the concept of convergence, recession, vanishing point, perspective, infinity which is inseparable from Gutenberg technology. For uniformity and repeatability are as basic to print as visuality to the phonetic alphabet.

Dantzig continues: "The importance of infinite processes for the practical exigencies of technical life can hardly be overemphasized. Practically all applications of arithmetic to geometry, mechanics, physics and even statistics involve these processes directly or indirectly. . . . Banish the infinite process, and mathematics pure and applied is reduced to the state in which it was known to the pre-Pythagoreans."[5] That is to say, without the minute segmentation, whether of alphabet or of the infinitesimal calculus, there can be no translation, no bridge from the tactile, resonating, tribal world, to the rational, flat, visual world.

Dantzig simply points out that number aided by infinite process can measure our world by translating visual, Euclidean space created by the phonetic alphabet back into the tactile modalities of touch and sound. One of the many prices we paid for abstracting ourselves from the tribal, multi-sensuous world was that we came to rely more and more on number to get us back into relation to that tribal world. It is not surprising therefore that number, the servant of letters, finally outgrew its master, civilization. For pushed all the way, number or tactile measurement gave us the new electric media which restore the resonating, tactile world as an immediate datum and all-embracing matrix of culture.

"Our notion of the length of an arc of a curve," says Dantzig,

> may serve as an illustration. The physical concept rests on that of a bent wire. We imagine that we have *straightened* the wire without *stretching* it; then the segment of the straight line will serve as the measure of the length of the arc. Now what do we mean by "without stretching?" We mean without a change in length. But this term implies that we already know something about the length of the arc. Such a formulation is obviously a *petitio principii* and could not serve as a mathematical definition.
>
> The alternative is to inscribe in the arc a sequence of rectilinear contours of an increasing number of sides. The sequence of these contours approaches a limit, and the length of the arc is defined as the limit of this sequence.[6]

Calculus, that is to say, is a means of translation of one kind of space into another – especially of visual into tactile and auditory fields of measurement.

> And what is true of the notion of length is true of areas, volumes, masses, moments, pressures, forces, stresses and strains, velocities, accelerations, etc., etc. All these notions were born in a *"linear," "rational"* world where nothing takes place but what is straight, flat, and uniform. Either, then, we must abandon these elementary rational notions – and this would mean a veritable revolution, so deeply are these concepts rooted in our minds; or we must adapt those rational notions to a world which is neither flat, nor straight, nor uniform.

> But how can the flat and the straight and the uniform be adapted to its very opposite, the skew and the curved and

non-uniform? Not by a finite number of steps, certainly! The miracle can be accomplished only by that miracle-maker, the *infinite*. Having determined to cling to the elementary rational notions, we have no other alternative than to regard the "curved" reality of our senses as the ultra-ultimate step in an infinite sequence of *flat* worlds which exist only in our imagination.[7]

The same navigational techniques of adaptation, compensation, and correction for distortion, which the mathematician provides for the sciences, the artist provides for sensibilities distorted by social technologies and media change.

Dantzig, of course, is wrong in supposing that there is anything elementary or uniquely rational about the fictions of Euclidean space reared upon our senses by the phonetic alphabet. The boast of Archimedes was fulfilled in the phonetic alphabet. The culture that uses it stands on the human eye and levers all the other senses into distorted configurations. Today, Archimedes can stand on the ear by radio or our tactile sense by television and enlarge the operation of these organs till they embrace the globe. But let us be quite clear that electric technology supplants and dissolves Euclidean rational space. As educators and responsible citizens, we have to inquire whether we choose to pay the price for a technological change which not only substitutes multiple spaces and times for our long-held Euclidean world, but which also pulls the rug out from under all the legal, political, and educational procedures of the past three thousand years of the Western world. These consequences ensue simply from stepping up information flow to the speed and level of the electronic. As is well-known in information theory, as information levels rise, not only one kind of natural resource becomes substitutable for another, but any subject-matter divisions also disappear. We are left confronting primal lines of force and development in a unified field. There are no subjects in a tribal, preliterate society, but there may be wisdom. There is no history, for all time is now. Even with manuscript culture there is little enough history. Manuscript moves information too slowly to permit the building of perspectives or to provide detailed pictorial background for men and events. Without the habit of a fixed point of view, which the printed book makes natural to a reader, there can be little or no perspective in historical attitudes. Today, we reconstruct past cultures and periods in all their living inter-relationships rather than trying to develop a single point of view. In the same way the anthropologist studies prehistoric or archaic man not with respect to his economic life, nor his art, nor his language, but with reference to all of these at once. As information levels rise, fixed point of view yields to inclusive multi-dimensional awareness. The poets and painters since Cézanne and Baudelaire have made us acquainted with these new patterns giving us our bearings and directions for the new age.

Perhaps a useful illustration of the same change that occurs in the development of historical attitudes and procedures can be found in the world of the press. The viewpoint press in which news is editorially digested, arranged, and directed to an audience underwent great change with the advent of the telegraph. When news pours into the office of a newspaper from many quarters and at high speed, it cannot be processed in the old viewpoint style. The writers of dispatches must adopt a neutral tone in order that the numerous items can be assembled in a single image. Nor is there any one point of view from which news from Tokyo, Peking, Formosa, Berlin, Moscow, New York, and London can be seen. The only course of the newspaper maker is to proceed on the mosaic basis. Under a single dateline he assembles the verbal and visual icons which the Teletype brings forth. Instead of addressing an audience from a policy point of view, he includes the audience as part of his inclusive mosaic image. "Human interest" records the moment when audience became the show. At this moment a radical political revolution was set in motion. This is the moment of birth of a mass medium. A mass medium is one in which the message is not directed *at* an audience but *through* an audience, as it were. The audience is both show and the message. Language is such a medium – one that includes all who use it as part of the medium itself. With telegraph, with the electric and the instantaneous, we encounter the same tribal inclusiveness, the same auditory and oral totality of field that is language.

This is but one of many instances of that principle of substitutability which comes into play as information levels rise very high. This principle now extends itself not only to our staples and natural resources but to our traditional disciplines, subject-matters, and even to our media of communication. But with the media, as with staples, if we are to permit one to do the work of another we must be fully acquainted with their properties. The telegraph press is not a substitute for the book nor for the viewpoint press. In order to adapt the telegraph to the newspaper or the telegraph press to the political ends of the viewpoint press, a great many adjustments would have had to be made in the educational process. These adjustments were never made. The result was a brainwashing of older political forms and the rise of new ones. Today, the notion that film, radio, or television can be allowed to convey the information of older subject-matters and disciplines is based on the assumption that the conveyors of information are neutral. An X-ray unit can get very hot but it is not at all suitable as a space heater. Our electric media are such potent units cast in the ridiculous role of space heaters. The printed book was long regarded as a cheap and vulgar form of manuscript. But this book did not extend the older forms of education to a wider public. It dissolved the dialogue and created wholly new patterns of political power and personal association. And our new science and mathematics not only dissolve Euclidean space but they re-create on a global scale the patterns of tribal life. We must grasp that electric speeds and modes of information movement create the conditions of the oral village for the entire planet. We must therefore be prepared to see the end of all of our concepts of the private person in relation to state and society. We must accept these consequences in the same spirit of somnambulist resignation with which we have urged our new technology to undertake the liquidation of all values and institutions which had been reared on the plateau of the older alphabetic and print technology.

E. H. Gombrich, in his recent *Art and Illusion*, regards cubism as "the most radical attempt to stamp out ambiguity and to enforce one reading of the picture – that of a man-made construction, a colored canvas."[8] Cubism, the means of seeing all aspects of an object from numerous points of view, at the same moment, is the near-equivalent of the telegraph press, which provides an inclusive global snapshot from hour to hour. Gombrich is right, however, in suggesting that when the ambiguities of perspective or the third dimension are pushed far enough they yield suddenly a reverse set of characteristics. Instead of pictorial space, we are suddenly confronted with formal space. Instead of a visual world that contains objects, we meet a world in which each object creates its own space, and imposes its own assumptions like a melody.

There is one passage in Gombrich's *Art and Illusion* which has instant appeal to a literary man. He is discussing the ambiguities of the third dimension as they are rendered in the Adelbert Ames perception laboratory. He wants to pin down the why of our desires to think of the third or perspective dimension as non-illusory: "It is important to be quite clear at this point wherein the illusion consists. It consists, I believe, in the conviction that there is only one way of interpreting the visual pattern in front of us."[9] This was also the most cherished illusion of the print reader. For reasons never yet investigated, the notion of the "one plain meaning" never "bugged" the manuscript reader, ancient or medieval. Possibly the higher definition of print created the expectation of exclusive rather than inclusive meaning. But it was only a generation ago that the literary world was startled by the rediscovery of multiple levels of statement in the simplest words and syntax. As we move deeper into the electronic galaxy the pressure to reconfigure age-old patterns in the alphabetic and Gutenberg galaxy becomes overwhelming.

It is therefore with ready understanding that we can nowadays confront the disturbance felt in the ancient world when the alphabet was new. The growth of the Euclidean fictions in the patterns of human sensibility were as upsetting then as the return of nuclear non-Euclidean modalities of experience today. Gombrich, writing of the rise of pictorial space and illusion in the sixth and fifth centuries B.C., says: "The very violence with which Plato denounces this trickery reminds us of the momentous fact that at the time he wrote, mimesis was a recent invention."[10]

And again: "There is finally the history of Greek painting, as we can follow it in painted pottery, which tells of the discovery of foreshortening and the conquest of space early in the fifth century and of light in the fourth."[11]

What we today can see very easily is that the departure of the Greek world into pictorial and Euclidean space was anything but natural. Preliterate, natural man then and now lived in a world of scheme which we encounter in child and primitive art. Such art allows no dominance to the eye. The multiple levels and modes of sound and tactility are favored in cave art above the visual. So it is with speech itself. But the reduction of speech to sight by the phonetic technology gave the eye an ascendancy over the other senses which is anything but natural to man. I am not making a value judgment. The natural may not be desirable at all. But the ascendancy of eye over the other senses gave us the miracle of mimesis, of foreshortening, and, eventually of perspective and vanishing point, which we have accepted as natural and rational for centuries. Such assumptions do not coincide with those of the electric media. The "message" of radio is that of a global tribal drum (the Hemingway Bell) awakening the most primal memories, disturbing literate visual man with a deep sense of his incompleteness and with a guilty awareness of selfish separateness and individualism. The message of radio to less- or newly literate societies is that of a high-definition mandate to step up tribal values to a new pitch of intensity. But the message of TV is tactility and the mandate to merge with the world of process. Concern with depth, inner and outer, and the craving for permanence and stability mark the emergent new TV generation. Whereas with the movies the audience became the camera, the insatiable eyes of the world, with TV the viewer is the screen. With the movie came the frantic extrovert generation of "I'll tell the world" fame. With TV we encounter an earnest introvert world whose aim is to be with it, disposed to a total commitment in depth which is also consciousness and articulation. For the obsession with process extends also to the process of making and knowing, of creativity and of order. And these tendencies are not of a kind to push the young towards a retreat to the womb.

Perhaps this note concerning our increasing drive towards understanding the creative process since TV may be the ideal point at which to take a stand. It might even be in order to Utopianize a bit. We members of a print culture are easily upset by new developments because we feel the need to fit new things into the old scheme. We tend to pictorialize and to classify data in visual patterns. When the data are not visual at all, they elude our schemes of order and cause confusion.

It is often said that the major development of this century has been the discovery of the night world of dreams. It is not the highly individualized world of mental gestures and collective choreography in which all human activities translate themselves into all other human actions, printing a book into bearing a baby, fighting a war into courting a woman. By day, we attempt originality; by night, plagiarism is forced upon us. By day, we are the bottom half of a double boiler. We are all steamed up but we don't know what's cooking.

In illuminating the night world, private and collective, Joyce in *Finnegans Wake* has only done what the electric light had done in abolishing the old divisions between night and day, and between inner and outer space, with respect to human work and play. As soon as the complementary dynamics of inner and outer, conscious and unconscious were displayed, it became easy to observe the operation of languages in shaping human assumptions, both sensuous and psychological. *Finnegans Wake* is an encyclopedia of lore concerning the origins and effects of words, of writing, of roads and bricks, of telegraph, radio, and television on the changing hues of the human spectrum.

Let us return for a moment to that increasing awareness of the dynamics of process and learning and creativity which I suggest gains new force from the subliminal patterns of the TV image. In his *Landmarks of Tomorrow*, Peter F. Drucker has pointed to Operations Research as "organized ignorance."[12] It is a procedure in tackling problems which resembles the "negative capability" of Keats – a sort of intellectual judo. Instead of straining all available effort on a visible goal or problem, let the solution come from the problem itself. If you can't keep the cow out of the garden, keep the garden out of the cow. A. N. Whitehead was fond of saying that the great discovery of the nineteenth century was not this or that invention but the discovery of the technique of invention. It is very simple, and was loudly proclaimed by Poe, Baudelaire, and Valéry, namely, begin with the solution to the problem, and then find out what steps lead to the solution. In other words, work backwards.

Such is Operations Research, in which metallurgic problems
are tackled by psychologists and historians but not by metal-
lurgists. For the expert knows too much about a problem in
advance. He sees why it is insoluble. But teams of intelligent
non-experts, not seeing the difficulties in advance, have, time
and again, won through, and at high speed. The new pattern in
management is small teams of men of varied competencies, not
the pyramid of job hierarchies. Just back from India, Dr.
Bernard Muller-Thym, one of the leading management consultants in the world, says that in the nuclear age, illiteracy is
India's biggest asset. He assumes the techniques of creativity of
Poe, Baudelaire, Joyce, and Operations Research.

Today, when the movement of information has itself
become by far the largest industry in the world, industry in
turn is ceasing to be mechanical and is becoming the scene
where computers and electronic tapes move the information
that is production. The workforce gradually retires from the
industrial scene. Where will it go? What will it do? The answer
to any problem is always in the problem. In our older economy
the problem had been to adjust the relations of supply and
demand. Today under conditions of instant information
movement, supply creates demand as readily as a floating
object displaces its weight in water. Information has itself long
been the largest consumer commodity. As the means of
moving information increase (and satellite broadcasting is a
startling development of this kind) there occurs a fluidity of
the categories of natural resources. For example, almost any
natural resource can be substituted for any other as levels of
information rise. The same is true of the categories of acquired
knowledge. Today, then, the globe is already a community of
learning. And the university, instead of being a place for the
processing of young minds, becomes the norm of human association. The human dialogue which occurs on the campus

assumes the character of a living model for states and communities globally.

Is not the emergence of a global community of learning a natural outcome of a world in which the production and transportation of commodities finally merges with the movement of information itself? Can we not derive a very great deal of hope and exhilaration from the fact that our highest technology not only derives from but directly nourishes the individual creative process and awareness? As Roy Harvey Pearce once put it to me, "Are we not about to discover the bases for human community by learning to live with and in terms of our media of communication after many centuries during which we lived in spite of and against the media?"

For these media, being our own faculties extended for the first time in human history to create a human sensorium outside us as well as within us, offer the immediate means of personal and social equilibrium if they are understood in their powers and influence. So considered, the values we cherish cannot but flourish more richly than in past societies. The act of cognition and recognition, which is the supremely human act, and which is yet common to all men, we outer or utter in art forms which have tended to be appropriated as consumer goods by various elites. Yet the inevitable mode of the electronic forms of information movement and of ensuing patterns of human expression and association are decisively producer-oriented. The audience is increasingly involved in the creative act, to the scandal of elites who have for so long assumed a consumer orientation to the arts. The dichotomies between elite and popular culture, like those between individual and society, are transitory discomforts stressed by a new technological galaxy as it invades and reconfigures another and older galaxy. If, during such a process of the print and literacy galaxy being invaded by an electric galaxy, we remain paralyzed by the oncoming imagery, we shall fail to see the new patterns of stress and tension assumed by the familiar components of the older galaxy. In this age, when not only non-print but non-verbal communication has become normal, we have got into a panic about literacy. Yet the very patterns which afflict the older modes of literacy have opened human awareness to the complexity and wealth of word formation and word action in the depths of our lives – such new knowledge has led to demonstrably better linguistic knowledge and to better reading habits. The printed word which once atomized society now acts to nucleate the social field. And so it is with the touchy question of individualism. These issues are altogether too massive for any individual's point of view to matter very much. But if we look closer at the organization world than Mr. Whyte has done, we shall find that the very improvements of communication, which at first appear to strangle the individual in industry, have also led to prime stress on creative autonomy in decision-making at all levels of organization. So painful at first is this for the older job-holder that it elicits such cries of anguish as "uncertainty economics." It is plain today that change is not only the constant in our society but that adjustment to change is quite impossible. We have no time to adjust and must substitute, instead, understanding of the process of change. But in the past, have not all free spirits deplored the ignominy of human adjustment and knuckling-under to the pressures of circumstance? We cannot knuckle under any longer, no matter how strong our private disposition to servility may be. The atomized and dehumanized teamwork patterns of the mechanical age are dissolving and departing. Most of us will miss them acutely, if only because they represented a world we had both molded and fought against. But in place of the old intransigence and rigidities we already discern new fluidity and flexibility such as will tax every individual's inner resources. We can see the new forms and the new demands on every hand. The higher education which has for so long been a luxury for elites is now a necessity of the most ordinary processes of production and planning. Shall we love the luxury the more when it is expected of each as a necessity? The answer is emphatically yes. For to exercise and to exchange knowledge is a delight which increases with its possession by our fellows. If the exercise and exchange of inexhaustible knowledge raises the quality of man, then in the electric age he will have his first universal opportunity to be richly human. But this extension of the human dialogue to embrace all men and all kinds of knowledge may appear to some to turn the globe itself into a single computer.

1. Bernard Berenson, *Florentine Painters of the Renaissance* (New York: Putnam's Sons, 1898), p. 4.

2. James Joyce, *Finnegans Wake* (London: Faber & Faber, 1939), p. 497.

3. Tobias Dantzig, *Number: The Language of Science* (New York: Doubleday, 1954), p. 139.

4. Ibid., p. 141.

5. Ibid., pp. 136-37.

6. Ibid., p. 137.

7. Ibid., pp. 137-38.

8. E. H. Gombrich, *Art and Illusion* (Princeton: Princeton University Press, 1960), p. 281.

9. Ibid., p. 249.

10. Ibid., p. 116.

11. Ibid., p. 117.

12. Peter F. Drucker, *Landmarks of Tomorrow* (New York: Harper & Brothers, 1959), p. 28.

■ ■ ■ ■ ■

THE COMMUNICATIONS REVOLUTION

Dale: We're going to discuss here for a while this whole matter of the communications revolution, and I suppose by this we mean the fact that in the last few decades we've had the great rise of the modern media – so-called radio, TV, motion pictures – and they've had a profound impact on our American civilization and, in fact, on all Western civilization. You did a nice job the other day, Mr. Seldes, delineating what's involved in all of this, and I think this might be a good way of beginning.

Seldes: I think that what you have in mind is the reference I made to the variety of scholars who are approaching the field, and I was trying and still am trying to assure myself we're more or less talking about the same subject, and I think we are in this sense that while you have an approach from a sociologist or a philologist or any number of ologists to the subject. . . . On one side, we're all trying to find out what is the nature of the new communications. What is the special nature of each one of the media such as the moving picture or broadcasting? What is the nature of all of them, all of the mass media, as we call them, put together? I think we then want to say what do all these media do to us, and then I put my own specialty in. What do we do about them? Now, I don't think I differ too much from you, Mr. McLuhan.

McLuhan: No, I've learned a great deal from your own work, Mr. Seldes, and you were in this field before any of us. And we've toiled along in your footsteps, as it were.

Seldes: You and I, however, share an enthusiasm for someone else, who is Harold Innis, the Canadian economist. Don't you think that he phrased the essence, really, of the revolution? You probably know the actual words that he used, of the connection between an enormous change in the means of communication and the change following in society.

Tyler: You know, McLuhan, you might do what I've known you to do, which is to characterize the former period as the print period and the present period as the electronic period and kind of point out what sharp distinctions you think are involved.

McLuhan: I think one of the things that happens when a new medium comes on the scene is you become aware of the basic characteristics of older media in a way that you were not when they were the only things around. I think we're becoming more aware now of what print is than we were before. Radio seems to have acquired more sense of its own identity since television, and movies likewise. So there's a great advantage in one way in this revolution brought on by a new medium in revealing some of the earlier features of older media, making them more intelligible and more useful, giving us more a sense of control over them.

Tyler: Needless to say, when we get into this electronic period we don't give up the print period.

McLuhan: No, it would seem natural that older forms are put to new uses and discover new roles. The book, for example, in our time has discovered many new functions that it never had fifty or a hundred years ago. It has become very powerfully directed toward teaching people how to learn other things besides books, how to learn arts of many kinds. The book has taken on a vast new function as a means of informing people, directing people's skills in many, many areas.

Dale: Mr. McLuhan, you've sometimes spoken about print as linear. Suzanne Langar in her book *Psychology in a New Key* speaks of discursive and non-discursive communication. How would you fit this into the scheme she has developed?

McLuhan: Well, I think people who are subjected to the arrangement of language visually in lines, highly sequential and precise and rigid, develop habits of arranging their lives, arranging their whole social existences which are very closely geared to these forms. They're not especially aware of this. Linearity, though, is not characteristic of radio or television or movies. And so we have been subjected to tremendous new forces, new influences, which have broken up the older habits acquired from the print world.

Seldes: Would you say we tend to think in straight lines until we break away from . . .

McLuhan: We still like to speak of following a person, or of drawing conclusions in lines. "And I don't follow you," or "I do follow you" sort of thing does suggest that we think of thought itself –

Seldes: But notice our contemporary phrase "I dig you."

McLuhan: This is closer, surely, to the medium of television which, incidentally, has had a strange effect on the young of driving them to the libraries to ask for fact books. The librarians report there is a tremendous new taste in the young for fact books, not fiction. Well, I mean I was thinking of the fact book as something you have to dig. You don't read it in a line, just a story level on a single plane. You have to dig a fact book. And the youngsters today dig their reading. They read in depth.

Tyler: But in a sense you're saying one of the characteristics of this revolution is that people are *with* things, that is to say we're communicating with them so much more rapidly that we are with the events rather than linearly learning about them afterward, event by event.

McLuhan: But not a merely descriptive or narrative relation or mere point-of-view relation. You see, if you have a point of view, you're not really with a situation. You have already abstracted an angle, an aspect as yours. In the new situation of being with, you don't have a point of view. You merely identify at all levels with your whole being.

Dale: Are you saying that photography and these new arts are more primitive, that we need to abstract less with these new media?

McLuhan: I think that the word *primitive* is misleading, perhaps, but it does suggest people who are less in the habit of abstracting single aspects, single levels, and so on, and who are accustomed to a more inclusive, totally sensuous commitment to situations.

Seldes: Then, McLuhan, picking up on Mr. Dale's *primitive*, indications are that we will have a less-sophisticated general public. You said when you're with it, you no longer need have a point of view. This is the complaint that's being made. People do not have their own points of view, their own specific individual approach to things. It has been said, indeed, the mass media have that effect because we all see the same things and so forth, but you are implying there's something in their own nature that reinforces this.

McLuhan: Yes.

Seldes: So that when I mentioned the Innis principle that there's a shift, there's a change, a social change that comes as a result of new means of communication, I find the implication of what you say is that it's not going to be a change in favor of the intellect per se. Well, will we have less intelligent people?

McLuhan: This "being with it," for example, is very character-istic of dialogue. In dialogue with somebody else, you are not maintaining a fixed point of view. You are interchanging, you are interpenetrating, modifying each, the point of view of each is modified by the dialogue itself. You do not hold a fixed point of view, which you do in an article in a magazine or in a book; you see, this is natural.

Seldes: You are saying that when I'm watching a television show, and I'm speaking now of an entertainment television show, that I'm in some way engaged in it and that I'm with it. I am engaging in a dialogue with whatever is going on in front of me?

McLuhan: The character of the television image, I think, fosters this kind of participation simply because it is a rather poor image, and it involves the viewer in a great deal of completion of the detail that is missing visually in that image. The act of seeing television is very much that of participation as in reading a detective story where you are very much with it precisely because you are not given very much narrative information. You have to fill it in.

Tyler: I'd like to suggest that this makes the job of the intellectual a good deal harder because he is with all of these events. He's got consciously to withdraw from them if he's going to have any time for reflection, for interpretation, and indeed for leadership. Otherwise, he's just a part of everything as it goes on and he never really has time to bring out what the essence of it is.

Dale: I wonder too if these photographic processes, these new processes, aren't leading people again back to print? I've heard people say, for example, take *Life* magazine, they started out with a tremendous amount of photographs and shifted to more print. Is it possible, Mr. Tyler, that radio or devices such as this have caused people to read better?

Tyler: I think certainly all these media interrelate to the extent to which they are used consciously. One of the difficulties, you mention radio, this is a hard one because radio is used so much today as a kind of background for other activities.

Seldes: Radio actually eliminates the noises or deadens the noise of the family so the children can study by it. I know this happened a long time ago, happened before I think I was aware of all the implications. I'd like McLuhan to go a little further because I was rather superficial about Innis's theory of change that follows.

McLuhan: His notion was that any change in handling information communication is bound to cause a great readjustment of all the social patterns, the educational patterns, the sources and conditions of political power. Public-opinion patterns will change. But he got onto that track rather interestingly. You see, as an economic historian he had been studying railways and the cod fisheries, the fur trade and the pulp and paper, and he moved then from staples as forms of shaping economic life to media as staples, and he began to study the new media as really basic economic resources. And much as, for example, cotton in the South has shaped a whole culture, now radio is shaping a global culture. It's global in the extent of its resource availability. So what would appear to be in the offing is a global culture conformable to a staple like radio as the Southern culture was conformable to cotton.

Seldes: Then he went on with the simple example that after print came in, the whole feudal system broke. Now what I'm worried about, and he didn't live long enough to predict any of it, is what is going to break? Where is the shift of power going to be at the present time?

McLuhan: The tremendous developments that we made in individual private habits of study, and of isolated effort, inner direction, and so on, these are likely to take the rap from media that are so inclusive of the whole of society and at all levels. Think of the tremendous shift in political power that is going on at this moment through the use of television in politics.

Seldes: Well, I really wonder about that. I have in the past thirty years observed only one demagogue using the visual medium, and that was Huey Long and movies. I'm told that he was very good on radio when he would start to talk and he would say, "Friends, I'm not going to say anything of importance for the next minute and a half. You get on the telephone and call up a couple of friends and get them to listen to this program." There was a very clever demagogic use but the only time he's scared me was when I saw him in movies and the newsreel, when I really felt he had the power.

McLuhan: McCarthy folded a week after he went on television. And if Huey Long had gone on television he would had been a flop at once. TV will not take a sharp character, a hot character. It's a cool medium, and our politics are being cooled off to the point of rigor mortis, according to many people. The nature of this medium which calls for so much participation does not give you a completed package, a completed image. You have to make your image as you go. Therefore, if the person who comes in front of the TV camera is already a very complete and classifiable type of person – a politician, a highly obvious doctor type, lawyer type – the medium rejects him because there's nothing left for the audience to view or to complete, and they say this guy's a phony. There's something wrong with this guy.

Tyler: I'd like to have you react to this notion that since the founding of our country, we've had the balance of power progressively go from a very small group that was voting and had property rights, and gradually spread to a larger and larger group. Now how does television fit into this? We have practically universal television as far as it being in homes. Does this mean that the power more and more is flowing to the popular group?

McLuhan: Yes, literally the participation of the whole population in the political process becomes very deep. It's no longer a question of assessing arguments, platforms, regional clashes, and so on. Everybody is with it – all age groups.

Seldes: Are you sure it's deep?

McLuhan: Yes, the issues are no longer given to you on single planes and single patterns. They are total.

Tyler: But this is a very good point about being deep. Is this actually a kind of a pseudo-event?

McLuhan: No.

Tyler: You think you are with reality but really you are having prepared for you on television those aspects they want you to see, which gives you a feeling of participation you may not really have.

McLuhan: The audience is making a new form of association among its own members. They are making a new reality, a new art form.

Seldes: You don't think we're learning more things superficially.

McLuhan: No, this is an age in which the New Criticism or psychology or anything else, the word used in all these forms is *depth*. Reading in depth. Psychology in depth. Everything now is in depth.

Seldes: These are the relatively few experts and outsiders.

McLuhan: No, no, no, no.

Seldes: Your implication now seems to be the opposite of what you were saying a few minutes ago. I'm not trying to trap you, Marshall. I'm trying to find out. Now, at one point, it's my understanding that the thing we need to be troubled about by these mass media is their creation of the non-individualistic person, the man who is with-it so he's with what everyone's doing, to use what is now a hackneyed word, the conformed person. On the other side, you are saying that television can be used to convey information and depth. The implication is that it can actually be used to make people think.

McLuhan: The forms of entertainment that work best on television, whether it's Paddy Chayefsky or even the Parr Show, are ones which admit of a great deal of casualness, in which people can be introduced and dialogued with in the presence of the camera at all sorts of levels of their lives. You capture them at all sorts of strange and offbeat moments of their existence. And this kind of probing and peeling off the superficial aspects of people is normal to this medium. It is a depth medium. The movie medium is, by comparison, very much a photographic slick-package medium which gives you a very highly defined and a very slick, complete package.

Tyler: But I still want to go back to a question I asked earlier, which is that with this feeling of being with everything, isn't there still the need, even more so for at least the intellectual, I would think, more and more for everyone to abstract himself from being with and have time for reflection and for becoming let's say an individual rather than simply a conformist mass man?

McLuhan: Well, intellectuals don't necessarily require privacy. They need the stimulus of intellectual conversation. But you can do a great deal of thinking in the course of a prolonged dialogue with able people.

Tyler: Not a dialogue with television.

Seldes: That is precisely where I'm with you. Can we really be thinking except the time when we stop looking?

McLuhan: I think it's a good time to bring up a point that when any new form comes into the foreground of things, we naturally look at it through the old stereos. We can't help that. This is normal, and we're still trying to see how will our previous forms of political and educational patterns persist under television. We're just trying to fit the old things into the new form, instead of asking what is the new form going to do to all the assumptions we had before. This is not good.

Seldes: I think all of us here learned to read and therefore protect books. But in the books we can read that when the book came in, it was denounced because people said how can you study from books without the authority of the teacher there? Now we say, how can you have television in the schoolroom without the authority of the book? And I see that. But I'm not sure there doesn't come a point when the vast amount of what we get from our new media doesn't really to an extent make it unnecessary to think deeply and maybe difficult to think. I'm dubious.

Dale: There's another point, too, here. I think with TV or film it's moving along at a continuous speed. In other words, you have no way of checking it. Now with a book you can check it. You can go back. You've got a kind of record. But we can't assume a person's going to get a videotape or is going to arrange to have a film re-run in a theater, so you don't have a chance to check certain of these new media. You just sort of have to take it as it comes along.

McLuhan: On the other hand, the degree to which, when reading, you are in the hands of an author, the degree to which he merely carries you along for a ride passively, has always been bypassed, not discussed, by book readers. They have the illusion of being engaged in great private intellectual discovery. Actually they're going for a ride.

Tyler: It's quite clear, gentlemen, we've raised a lot of questions and we're not going to answer them in one short discussion.

McLuhan: Surely it's very valuable to get these things out there for inspection, to know what are the relevant issues.

Cybernetics and
Human Culture
(1964)

In 1962, McLuhan published The Gutenberg Galaxy: The Making of Typographical Man, *his study of the effects of typography on the Western individual and society, and two years later he published* Understanding Media: The Extensions of Man, *which details his revolutionary ideas about the process of communication and the social change brought about by the new electric media. These two books brought McLuhan international acclaim as the world's foremost authority on communications. In 1963, he was appointed by the president of the University of Toronto to create a new Centre for Culture and Technology to study the psychic and social consequences of technology and media.*

On November 20, 1964, McLuhan was invited to attend the Symposium on Cybernetics and Society, part of the 175th Anniversary Year Program of Georgetown University. The conference, held in Washington under the joint sponsorship of Georgetown University, American University, and George Washington University, and set up in cooperation with the newly formed American Society of Cybernetics, had as its general theme "The Implications of Cybernetics for Social Order and Individual Fulfillment."

In his lecture titled "Cybernetics and Human Culture," McLuhan starkly contrasts the new electronic age of cybernation

with the old visual/mechanical age. "By awakening to the signifi-
cance of electronic feedback, we have become intensely aware of
the meaning and effects of our actions after centuries of compara-
tive heedlessness and non-involvement."

■ ■ ■ ■ ■

Today young people are in the habit of saying, "Humor is not
cool." The old-fashioned joke with its story line has given way
to the conundrum, e.g.: "What is purple and hums?" Answer:
"An electric grape." "Why does it hum?" "Because it doesn't
know the words!" This kind of joke is involving. It requires
the participation of the hearer. The old-fashioned joke, on the
other hand, permitted us to be detached, and to laugh *at* things.
The new kind of joke is a gestalt or configuration in the style of
set theory. The old-fashioned story is a narrative with a point
of view. This helps to explain a strange aspect of humor raised
by Steve Allen. In his book *The Funny Men*, Steve Allen suggests
that the funnyman is a man with a grievance. In Canada at
present, the Quebec separatists are a people with a grievance. A
whole stock of stories has come into existence in connection
with their grievances. For example, there is the story of the
mouse that was being chased by the cat. The mouse finally discovered a spot under the floor to hide in. After a while it heard a
strange "Arf! Arf! Bow! Wow!" sort of sound and realized that
the house dog must have come along and chased the cat away.
So the mouse popped up, and the cat grabbed it. As the cat
chewed it down, the cat said, "You know, it pays to be bilingual!"
     The story line as a means of organizing data has tended
to disappear in many of the arts. In poetry it ended with

Rimbaud, in painting it ended with abstract art, and in the movie it ends with Bergman and Fellini. One way of describing our situation in the electronic age is to say that we have come to the end of the neolithic age. The neolithic age brought in thousands of years ago a new pattern of work and organization. It represented a transition from the age of the nomadic hunter and the food-gatherer to the age of the sedentary and agrarian man. The neolithic time began with the specializing of human work and action that gave us our great handicrafts, including script and writing, whether on stone or papyrus. Not until script, around 3000 B.C., did man begin the first enclosures of space that we call architecture. There is no architecture known in any culture earlier than script. The reasons for this are very instructive. Man's orientation to space before writing is non-specialist. His caves are scooped-out space. His wigwams are wraparound, or proprioceptive, space, not too distant from the Volkswagen and the space capsule! The igloo and the pueblo are not enclosed space; they are plastically modeled forms of space, very close to sculpture.

Sculpture itself, which today is manifesting such vigor and development, is a kind of spatial organization that deserves close attention. Sculpture does not enclose space. Neither is it contained in any space. Rather, it models or shapes space. It resonates. In his *Experiments in Hearing*, Georg von Békésy found it expedient to explain the nature of sound and of auditory space by appealing to the example of Persian wall painting. The world of the flat iconic image, he points out, is a much better guide to the world of sound than three-dimensional and pictorial art. The flat iconic forms of art have much in common with acoustic or resonating space. Pictorial three-dimensional art has little in common with acoustic space because it selects a single moment in the life of a form, whereas the flat iconic image gives an integral bounding line or contour that represents not one moment or one aspect of a form, but offers instead an inclusive integral pattern. This is a mysterious matter to highly visual and literate people who associate visual organization of experience with the real world and who say, "Seeing is believing." Yet this strange gap between the specialist, visual world and the integral, auditory world needs to be understood today above all, for it contains the key to an understanding of what automation and cybernetics imply.

To anticipate a bit, and to capsulate a good deal, let me suggest that cybernation has much in common with the acoustic world and very little in common with the visual world. If we speak in configurational terms, cybernation tends to restore the integral and inclusive patterns of work and learning that had characterized the age of the hunter and the food-gatherer but tended to fade with the rise of the neolithic or specialist revolution in human work and activity. Paradoxically, the electronic age of cybernation is unifying and integrating, whereas the mechanical age had been fragmenting and dissociating.

Today, at the end of the neolithic age, we have the Bomb as environment. The Bomb is not a gimmick or a gadget. It is not something that has been inserted in the military establishment any more than automation is something that is now being inserted into the industrial establishment. The Bomb, like automation, is a new environment consisting of a network of information and feedback loops. In moving from the mechanical to the electronic age, we move from the world of the wheel to the world of the circuit. And where the wheel was a fragmenting environment, the circuit is an integrating environmental process. The Bomb, as pure information, consists of higher learning. It is, as it were, the extension division of the modern university in its highest research areas, creating a very tight environment indeed.

The Bomb takes over all earlier technology and organization. It also makes us vividly aware of all human cultures as responsive cybernetic systems. We are never made aware of our environment until it becomes the content of a new environment. The culture in which a man lives consists of structures based on ground rules of which we are mysteriously unconscious. (This is a matter that has been eloquently demonstrated in *The Silent Language* by Edward T. Hall.) But any change in the ground rules of a culture nonetheless modifies the total structure, and cybernation, far more than railway or airplane, speeds up information movement within a culture, effecting total change in perception and outlook and social organization.

In moving from the neolithic age to the electronic age, we move from the mode of the wheel to the mode of the circuit, from the lineal single-plane organization of experience to the pattern of feedback and circuitry and involvement. During the many centuries of specialist technology, man cultivated habits of detachment and indifference to the social consequences of his new specialist technologies. In the age of circuitry, the consequences of any action occur at the same time as the action. Thus we now experience a growing need to build the very consequences of our programs into the original design and to put the consumer into the production process. By awakening to the significance of electronic feedback, we have become intensely aware of the meaning and effects of our actions after centuries of comparative heedlessness and non-involvement.

Another way of looking at our situation today in the age of cybernation and information machines is to say that from the time of the origin of script and wheel, men have been engaged in extending their bodies technologically. They have created instruments that simulated and exaggerated and fragmented our various physical powers for the exertion of force, for the recording of data, and for the speeding of action and association. With the advent of electromagnetism, a totally new organic principle came into play. Electricity made possible the extension of the human nervous system as a new social environment. In 1844, Søren Kierkegaard published his *Concept of Dread*. This was the first year of the commercial telegraph, and Kierkegaard manifested clairvoyant awareness of its implications for man. The artist tends to be a man who is fully aware of the environmental. The artist has been called the "antennae" of the race. The artistic conscience is focused on the psychic and social implications of technology. The artist builds models of the new environments and new social lives that are the hidden potential of new technology.

In *Fortune* magazine, August 1964, Tom Alexander wrote an article entitled "The Wild Birds Find a Corporate Roost." The "Wild Birds" are science-fiction writers retained by big business to invent new environments for new technology. Big business wants to know what kind of world it will have created for itself in ten years or so. That is to say, the big enterprises have become aware that their technological innovations tend to create new environments for enterprise and for bureaucracy. Yet these environments are almost imperceptible except to the artist. If, in fact, the businessman had perceptions trained to read the language of the arts, he would be able to foresee not ten but fifty years ahead in all fields of education, government, and merchandising. It is one of the ironies of the electronic age that the businessman must become alert and highly trained in the world of the arts. It is one of the mysteries of cybernation that it is forever challenged by the need to simulate consciousness. In fact, it will be limited to simulating specialist activities of the mind for some time to come. In the same way, our technologies have for thousands of years simulated not the body, but fragments thereof. It was in the city alone that the image of the human body as a unity became manifest.

In *Preface to Plato*, Eric Havelock traces the changeover from tribal to civilized society. Before the environmental pervasiveness of writing occurred in the fifth century B.C., Greece had educated its young by having them memorize the poets. It was an education dedicated to operational and prudential wisdom. It is sufficiently manifest in the *Odyssey* of Homer. The hero of that poem, the wily Ulysses, is called *Polumetis*, the man of many devices. The poets provided endless practical illustrations of how to conduct oneself in the varied contingencies of daily life. This type of education Havelock very fittingly calls the tribal encyclopedia. Those undergoing this type of education were expected to know all things whatever, in heaven or on earth. Moreover, they were expected to share this wisdom with all members of the tribe, much as educated English people today are expected to know *Alice in Wonderland*. It was, therefore, a considerable revelation when writing came to detribalize and to individualize man. In creating the detribalized individual, phonetic literacy created the need for a new educational program. Plato, says Havelock, was the first to tackle this problem directly. Plato came up with a spectacular strategy. Instead of the tribal encyclopedia, he advocated classified data. Instead of corporate wisdom, he taught analytic procedures. Instead of the resonating tribal wisdom and energy, Plato proposed a visual order of ideas for learning and organization. The fascinating relevance of Havelock's book for us today is that we seem to be playing that Platonic tape backwards. Cybernation seems to be taking us out of the visual world of classified data back into the tribal world of integral patterns and corporate awareness. In the same way, the electronic age seems to be abolishing the fragmented and specialist form of work called jobs and restoring us to the non-specialized and highly involved form of human dedication called roles. We seem to be moving from the age of specialism to the age of comprehensive involvement.

Since this is a very confusing and even terrifying reversal in human affairs, it may be helpful to take a second look at the general pattern of development. We may take some consolation from the anecdote of a clerk at a toy counter. When he saw a customer curiously scrutinizing one of the toys, he spoke up. He said, "Madam, I can recommend that toy. It will help your child to adjust to the modern world. You see, no matter how you put it together, it's wrong!"

In approaching the matter of the significance of cybernation as an environment of information, it is helpful to consider the nature and function of other environments created by other extensions of the human organism. For example, clothing as an immediate extension of our skin serves the function of energy control and energy channel. An unclad population even in a warm climate eats 40 per cent more than a clad population. Clothing serves, that is, as a conserver of energy for doing tasks that the unclad could not undertake. The unclad man in the jungle after twenty-four hours without food and water is in dire straits. The heavily clad Eskimo at sixty degrees below zero can go for days without food. Clothing, as a technology, is a store of energy. It enables man to specialize. The consequences of clothes in terms of changing sensory organization and perception are very far-reaching indeed.

One of the most fantastic examples of the consequences of seemingly minor technological change is described by Lynn White in *Medieval Technology and Social Change*. His first chapter concerns the stirrup, the extension of the foot. The stirrup was unknown to the Greeks and Romans and came into the early medieval world from the East. It enabled men to wear heavy armor on horseback. Men became tanks. It became mandatory to have this equipment; yet one suit of armor required the skilled labor of one man for an entire year. The landholding system did not permit small farmers to pay for such equipment. To finance the production of armor so necessary to the needs of social elites, it became expedient to reorganize the entire landholding system. The feudal system came into existence to pay for heavy armor. When the new technology of gunpowder appeared, it blew the armor right off the backs of the knights. Gunpowder changed the ground rules of the feudal system as drastically as the stirrup had changed the ground rules of the ancient economy. It was as democratic as print.

The extension of the nervous system in electronic technology is a revolution many times greater in magnitude than such petty extensions as sword, and pen, and wheel. The consequences will be accordingly greater. At the present time, one area in which we daily observe the confusion resulting from sudden change of environmental factors is that of the educational dropout. Today, the ordinary child lives in an electronic environment; he lives in a world of information overload. From infancy he is confronted with the television image, with its braille-like texture and profoundly involving character. It is typical of our retrospective orientation and our inveterate habit of looking at the new through the spectacles of the old that we should imagine television to be an extension of our visual powers. It is much more an extension of the integrating sense of active touch. Any moment of television provides more data than could be recorded in a dozen pages of prose. The next moment provides more pages of prose. The children, so accustomed to a Niagara of data in their ordinary environments, are introduced to nineteenth-century classrooms and curricula, where data flow is not only small in quantity but fragmented in pattern. The subjects are unrelated. The environmental clash can nullify motivation in learning.

Dropouts are often the brightest people in the class. When asked what they would like to do, they often say, "I would like to teach." This really makes sense. They are saying that they would rather be involved in the creative processes of production than in the consumer processes of sopping up packaged data. Our classrooms and our curricula are still modeled on the old industrial environment. They have not come to terms with the electronic age and feedback. What is indicated for the new learning procedures is not the absorption of classified and fragmented data, but pattern recognition with all that that implies of grasping interrelationships. We are actually living out the paradox of having provided cities that are more potent teaching machines than our formal educational system. The environment itself has become richer. We seem to be approaching the age when we shall program the environment instead of the curriculum. This possibility was foreshadowed in the famous Hawthorne experiment.

Elton Mayo's group found at the Hawthorne plant that whether they varied working conditions in the direction of the agreeable or the disagreeable, more and better work was turned out. They concluded that observation and testing, that is, an involved environment as such, tended to change the entire work situation. They had discovered that the prepared environment for learning and work must be ideally programmed for new perception and discovery. The workers at the Hawthorne plant were not merely being observed. They were sharing in the process of discovery. The classroom and the curriculum of the future will have to have this built-in pattern of discovery in order to match the potential of improved information movement. The world of cybernation offers the immediate possibility of programming all education for discovery instead of for instruction and data input. This was the great discovery of Maria Montessori, who found that the prepared environment worked wonders far beyond the level of the prepared curriculum.

One of the misconceptions attending the onset of cybernation and automation is the fear of centralism. Indeed, on all hands automation is greeted as a further development of the mechanical age. In fact, automation abolishes the patterns and procedures of the mechanical age, though at first, like the horseless carriages with large buggy-whip holders, a new technology is set to perform the old tasks that are quite unsuited to it. Cybernation in effect means a new world of autonomy and decentralism in all human affairs. This appears obvious in so basic a matter as electric light and power, but also applies at all levels. For example, the effect of Telstar, when it is fully operative, will be to supplant the centralism of present broadcasting networks. Instead of extending a common pattern to a whole range of human affairs, the tendency of cybernation is toward the custom-built in production and toward autonomy and depth in learning. It would be easy to illustrate these patterns of development from the world of poetry and painting and architecture in our time. My topic suggests that I should relate it to the world of ethics and religion. One need mention only the ecumenical movement or the liturgical movement of our time to get one's cultural bearings in these matters. Both of these movements have in common an emphasis on the pluralistic and stress on participation and involvement.

To many people these new patterns seem to threaten the very structure of personal identity. For centuries we have been defining the nature of the self by separateness and non-participation, by exclusiveness rather than inclusiveness. It is true that the electronic age, by creating instant involvement of each of us in all people, has begun to repattern the very nature of identity. All the philosophers and artists of the past century have been at grips with this problem. But whereas before the problem of identity had been one of meagerness and poverty, it has now become the problem of abundance and superfluity. We are individually overwhelmed by corporate consciousness and by the inclusive experience of mankind both past and present.

It would be a comic irony if men proved unable to cope with abundance and riches in both the economic and psychic order. It is not likely to happen. The most persistent habits of penury are bound to yield before the onslaught of largesse and abundant life.

The Future of Man
in the Electric Age
(1965)

When McLuhan's most recent book, Understanding Media: The Extensions of Man, *was published in England in fall 1964, his theories about the mass media immediately caught the attention of the British press. On January 24, 1965, he became the subject of the British Broadcasting Corporation's program* Monitor *with the literary critic Frank Kermode as interviewer.*

Kermode is clearly familiar with McLuhan's ideas and confronts him with the notion that his vision of the future is that of a big brother image. McLuhan retorts: "We have never stopped interfering drastically with ourselves by every technology we could latch on to. We have absolutely disrupted our lives over and over again." McLuhan also describes aspects of his theory that turn common sense upside down: why television, which seems so visual, is a tactile medium; why television, which seems so shallow, is a depth medium; and why television, which seems so ploddingly literal-minded, is a mythic medium.

Kermode: We're going to be using the word *technology* probably rather a lot in this conversation. So, could we start by asking you exactly what you mean in the context of your argument by this word?

McLuhan: I try not to have any private meanings, Frank, but I think of technologies as extensions of our own bodies, of our own faculties, whether clothing, housing, and the more familiar kinds of technologies like wheels, stirrups, extensions of various parts of the body. The need to amplify the human powers in order to cope with various environments brings on these extensions, whether of tools or furniture. These amplifications of our powers, sorts of deifications of man, I think of as technologies.

Kermode: Yes, and by developing them we alter the whole pattern of our senses, you argue.

McLuhan: They create environments. Every technology at once rearranges patterns of human association and, in effect, really creates a new environment which is perhaps most felt although not most noticed in changing sensory ratios and sensory patterns.

Kermode: Can I put this to you? In a sense you have been an historian as you've gone about your work, and let's talk first a little about your book, *The Gutenberg Galaxy*, where you argue that for a long time, without actually understanding it, we've been living in a culture in which our whole way of looking at the world has been determined by typography, by the successiveness of print, and so on. Would you like to enlarge on that a bit?

McLuhan: I remember I decided to write that book when I came across a piece by J. C. Carothers, a psychiatrist, on the

African mind in health and disease describing the effects of the printed word on the African populations. It startled me and decided me to plunge in. But we have a better opportunity to see our old technologies when they confront other populations elsewhere in the world. The effects they have on most people are so startling and so sudden that we have an opportunity to see what happened to us over many centuries.

Kermode: Yes, which we couldn't see because we're inside the system. Don't you say that what happened was that we got used to having our information processed as it is in print, that is to say set out successively? Whereas at the root of your thoughts, perhaps, there is the view that we can see the world as an image instantaneously, but that we've chosen under the pressure of a technology to set it out successively like a block of print?

McLuhan: Every technology has its own ground rules, as it were. It decides all sorts of arrangements in other spheres. The effect of script and the ability to make inventories and collect and store data changed many social habits and processes back as early as 3000 B.C. However, that's about as early as scripts began. The effects of rearranging one's experience, organizing one's experience by these new extensions of our powers, are quite unexpected. I think we needn't plunge in at this time, but I think one of the very definite effects of scripting was the rise of what we call architecture. It made possible, instead of just scooped-out cave spaces or wraparound wigwams and igloos, enclosed space.

Kermode: Now, you're talking about script rather than print.

McLuhan: The effect of writing as a way of organizing experience stepped up the visual factor until things like architecture became possible.

Kermode: So could we say this? Would this represent your view that if you go back far enough, to a kind of primitive culture, you'll have a state of affairs where our view of the world is not affected by the way in which we process it technologically? In other words, when you begin to write you immediately begin to get a degree of successiveness, and then when you begin to print you get an increased and more private degree because people read quietly instead of . . .

McLuhan: I think, Frank, that perhaps one way of putting it is to say that writing represents a high degree of specializing of our powers. Compared to preliterate societies, there's a considerable concentration on one faculty when you develop a skill like scripting.

Kermode: Well, this is the visual, what you call the visual sense.

McLuhan: Yes, this is a highly specialized stress compared to anything in ordinary oral societies. There have been many studies made of this in various ways, but in our own Western world the rise of the phonetic alphabet seems to have had much to do with Platonic culture and the ordering of experience in terms of ideas, the classifying of data and experience by ideas.

Kermode: You mean that sight has become the pre-eminent sense as it was with Plato, and it went on being so in so-called civilized as opposed to primitive societies.

McLuhan: Increasingly so.

Kermode: And climaxed with the invention of printing.

McLuhan: Printing stepped it up to a considerable pitch, yes.

Kermode: How would you describe the impact of the invention of the printing press? Give us some instances of what happened as a consequence of it.

McLuhan: It created almost overnight what we call a nationalism, what in effect was a public. The old manuscript forms were not sufficiently powerful instruments of technology to create publics in the sense that print was able to do – unified, homogeneous, reading publics.

Kermode: Yet somehow private?

McLuhan: Very highly fragmented. The fragmented, private outlook of literate man is very much related to typography and tends to disappear, of course, with changes in technology such as the electronic.

Kermode: What about the Reformation? You argue in your book *The Gutenberg Galaxy* that the Reformation is a kind of consequence of our moving into a typographical phase of history.

McLuhan: I don't think I stress that. I left that more or less to be implied. Everything that we prize in our Western world in matters of individualism, separatism, and of unique point of view and private judgement, all those factors are highly favored by the printed word, and not really favored by other forms of culture like radio or earlier even by manuscript. But this stepping-up of the fragmented, the individual, the private judgement, the point of view, in fact our whole vocabularies underwent huge change with the arrival of such technology.

Kermode: Would it be fair to say, or is this pushing your opinion too far, that the concept of the liberty of the individual, of the freedom of speech, if you like, is essentially a typographical matter which could well disappear in the other kind of culture that we're now moving into?

McLuhan: It could indeed if it hasn't already. The whole stress on point of view also creates the image of the importance of expression instead of the importance of belonging and being involved in a deep role in society. The need for private expression comes in with the technical possibility of extending one's voice or one's outlook into the community.

Kermode: So a whole phase of art, in fact of poetry, belongs to this typographically determined culture and may end with it, although, in fact, action painting, for example, is clearly a hangover from that as well as moving into the new kinds of –

McLuhan: It straddles, I think. It has much of both worlds.

Kermode: This concept of straddling is very important to you, isn't it?

McLuhan: Yes.

Kermode: You mention that it is at the moment of interpenetration of two different technological epochs that certain kinds of radical self-consciousness become possible. That's probably not quite the way you put it.

McLuhan: A new technology tends to take as its content the old technology so that the new technology tends to flood any given present with archaism as indeed Plato did with the dialogue. It was the old oral culture.

Kermode: There's a sense, of course, in which all civilized institutions are archaic like the law, isn't there? I mean it lives on its archaism.

McLuhan: When print was new, it flooded the Renaissance with medieval materials. In fact, the Renaissance had very little else.

Kermode: And with ancient materials.

McLuhan: Ancient but mainly medieval. And when the mechanical technology and later on the industrial was new, it surrounded the old agrarian world and turned arts and crafts and nature itself into kind of art forms – this tendency of a new technology to envelop an old one and to upgrade them into our awareness, to make us much more aware of what had been there all along.

Kermode: And that's the value of the moment of interpenetration?

McLuhan: I'm not sure.

Kermode: Let's put it this way. We've now, I think, got some idea of what you meant by Gutenberg technology, by a phase of history which is determined by the invention of printing. I think we might emphasize that it does involve to a marked extent the treatment of information as sequential, as not coming instantaneously, but as set out in the stereotypes of the phonetic alphabet. Could I ask you now about the technology which on your view is superseding it and which is having its own effect on our lives comparable with, though of course entirely different in kind to, the Gutenberg technology?

McLuhan: Gutenberg technology was mechanical to an extreme degree. In fact, it originated a good deal of the later mechanical revolution assembly-line style and the fragmentation of the operations and functions as the very rationale of industrialization. This fragmentation had begun much earlier after the hunter and the food-gatherers with neolithic man. I suppose in an extreme way one might say that Gutenberg was the last phase of the neolithic revolution. Gutenberg plus the industrial revolution that followed was a pushing of specialism that came in with the neolithic man, the agrarian revolution, a pushing of specialism all the way. And then suddenly we encounter the electric or electromagnetism, which seems to have a totally different principle, is, some people feel, an extension of our nervous system, not an extension merely of our bodies. If the wheel is an extension of feet, and tools of hands, back, arms, electromagnetism seems to be in its technological manifestations an extension of our nerves and becomes mainly an information system. It is, above all, a feedback or looped system.

Kermode: Yes. Now this is something that began to move in on us with the invention of the telegraph. We're now quite far advanced into what you call the electric galaxy.

McLuhan: Over a century in. But the peculiarity is, you see, that after the age of the wheel, you suddenly encounter the age of the circuit. The wheel pushed to an extreme suddenly acquires opposite characteristics. This seems to happen with a good many technologies that if they get pushed to a very distant point, they reverse their characteristics. The wheel reversed its characteristics when it became

an electric circuit or loop, and the feedback in that loop
system has a completely different set of effects on psyche
and society from any effects that the old mechanical technologies had.

Kermode: What difference is the electric technology making to
our interest in content, in what the medium actually says?

McLuhan: Very complex business, I'm sure, but one of the
effects of switching over to circuitry from mechanical moving
parts and wheels is an enormous increase in the amount of
information that is moving. You cannot cope with vast
amounts of information in the old fragmentary classified patterns. You tend to go looking for mythic and structural forms
in order to manage such complex data moving at very high
speeds. So the electrical engineers often speak of pattern
recognition as a normal need of people processing data electrically and by computers, a need for pattern recognition. It's
a need which the poets foresaw a century ago in their drive
back to mythic forms of organizing experience.

Kermode: Yes, and what you're saying is that there is a kind of
built-in primitivism in this antithetical technological system
that we've got now.

McLuhan: Yes, in the midst of complaining about our
superficiality, our lack of integral qualities, we've suddenly
found ourselves in a super-primitive swamp of integral,
involved mankind. All the propaganda for primitivism
seems to have come home to roost in a spectacular way with
very large quantities of it being provided by the new technology. We're living mythically now. We've continued to
think in the old rational patterns of the older technologies.
But we are suddenly forced to live in such complex and
compressed and high-speed systems that we inevitably
switch into mythic patterns.

Kermode: I want to try and get down crudely to the principal
differences between what you call typographical and what

you call electric culture. Now the first is visual and the second is aural.

McLuhan: We haven't gotten into that visual one. The visual sense is becoming more and more noticed, has properties that are quite unique to the visual sense and not characteristic of the other senses. I think it was Alex Leighton who said, "To the blind, all things are sudden." Without the sense of sight there is an absence of continuity and uniformity and connectiveness in the ordinary data of experience. And there is this kind of expectation of living in a space such as we sit in or as most people live in, in which we believe that all parts of it are uniform and homogeneous, and that one space is very much like another space.

Kermode: How would this come about as a result of typography?

McLuhan: Or as a result of literacy. As early as Euclid, the expectation that space is uniform and continuous and connected became the very basis of Greek science. But to relativity and to nuclear physics and electronic man, there is no such space available.

Kermode: Yes, what about time?

McLuhan: Similarly, time takes on many of the characteristics of a literate culture, of a uniformly homogenized and connected visual culture as well as space. We extend the cultural pattern to all facts of our existence and experience.

Kermode: But it would be very hard to show that there is another culture, I think, in which time was not thought of roughly as we think of it. Otherwise there would be no way of our communicating with primitive cultures.

McLuhan: You are probably familiar with the ancient Chinese efforts to measure time by the sense of smell.

Kermode: I'm not, no.

McLuhan: "The Scent of Time" is a fascinating recent study on that subject.[1] But the idea of time as a continuum, as a uniform and connected continuum, is not at all present to the oral societies. This is a particular allusion of visual and literate man.

Kermode: Don't we have this in our own traditions, the distinction between merely successive time and time thought of as seasons? And I suppose this is always being brought back to us by accounts of mystical experiences and so on. Isn't this continuous? This happens at all periods, the sense of intemporality in certain experiences is something we know about.

McLuhan: The experience of time has always been discontinuous. The ways of measuring it took on this visual character clockwise with literacy and it's no longer satisfactory for many scientific purposes.

Kermode: But how would the arrival of the electric technology modify our views of time, for example?

McLuhan: I think it is well enough manifested in Lewis Carroll, where both time and space are suddenly subjected to total discontinuity and non-conformities and disconnections.

Kermode: We've always been aware of this in dreams, of course, haven't we?

McLuhan: Yes, Lewis Carroll or Charles Dodgson was a mathematician who had become quite fascinated by non-Euclidean spaces and non-Euclidean geometries, and he has transferred those to the nursery world, but, in effect, created a very avant-garde and symbolist treatise by so doing. So our world beginning as early, in fact, as Rimbaud, our whole environmental world electromagnetically moved into a world of discontinuity and disconnection and unique spaces rather than homogeneous spaces.

Kermode: Now this way of talking really depends, doesn't it, on what you said before, namely, that these changes occur without people recognizing them until we reach a certain point. A lot of people presumably don't know that these changes are happening to them. How will they recognize them? How, in their ordinary use of the media, the newspaper, the radio, and television, how will they recognize these changes?

McLuhan: They frequently encounter one medium bumping into another medium. When the soundtrack is put on movies, when the radio goes around the picture, all sorts of strange things happen to make them aware. At the present time, the TV uses as its content the film. Movies have become the content of TV. We're totally unaware of TV as a form, tremendously aware of the content, which is the movie, much more so than when the movie was an environment. When the movie was environmental, it was relatively imperceptible. Now that it has become the content of TV, we can see it clearly.

Kermode: Well, here we are, a couple of archaic, literate men, Gutenberg men, talking on the television. What is the audience getting from this? Is it listening to what we're saying? Or is it feeling the impact of a new electric medium?

McLuhan: There is a book called *Is Anybody Listening* by William H. Whyte Jr. It's what worries the advertising men a great deal. The idea of feedback, of being involved, of participating in one's own audience participation is a natural product of circuitry. Everything under electric conditions is looped. You become folded over into yourself. Your image of yourself changes completely.

Kermode: In your other book, the more recent one, *Understanding Media*, where you go into all this, you use a kind of slogan. I think the expression is "the medium is the message." Would you like to illuminate on that?

McLuhan: I think it is more satisfactory to say that any medium, be it radio or be it wheel, tends to create a completely new human environment. The human environment as such tends to have a kind of invisible character about it.

The unawareness of the environmental is compensated for by some attention to the content of the environment. The environment as merely a set of ground rules, as a kind of overall enveloping force, gets very little recognition as a form except from the artist. I think our arts, if you look at them in this connection, do throw quite a lot of light on environments. The artist is usually engaged in somewhat excitedly explaining to people the character of new environments and new strategies of culture necessary to cope with them. Blake is an extreme case of a man who was absolutely panicked by the kind of new environment that he saw forming around him under the auspices of Newton and Locke and industrialism. He thought it was going to smash the unity of the imaginative and sensory life all to bits. What he was insisting upon in his own lifetime became quite a popular and widespread movement later on.

Kermode: Can I return to television? Because here we are, and whoever is listening to us is also undergoing the impact of television at the moment. On your view, they're all deceiving themselves in so far as they're paying attention to what we're saying because what's going on is a medium which is in itself the image that they ought to be concerning themselves with.

McLuhan: The medium of television has many characteristics which have been unheeded. Mostly it is seen under the aspect of movie form. The TV camera does not have a shutter, does not take pictures. It handles, it picks up, as radio picks up, the TV camera picks up its environment, handles it, scans it, and the effect of the TV image is iconic in the sense that it shapes things by contours rather than by little snapshots.

Kermode: This is one of the words that you use a good deal – *iconic*. I think we had better be clear about what you mean.

McLuhan: I think of it again, to tie in with Blake, his whole insistence upon the engraved, the highly patterned and highly sculptured forms and images; the iconic in that sense is very low in visual quality, very high in tactual quality, active touch, not cutaneous but "active touch," as the psychologists say.

Kermode: You call television a tactile medium.

McLuhan: Iconic medium having much in common with the cartoon for which it is ideally suited, much more well-suited than for pictures.

Kermode: Let me put a couple of possible objections to that. First, in a sense, when you look at a picture, whatever kind of picture it is, a temporal process is involved because you scan this picture, don't you? It doesn't matter whether it is a Mantegna or *Coronation Street*. You scan it. The television camera also does this as I understand it, not that I do understand it, but it proceeds along its lines. It's a moving point scanning very quickly. Anyway, it's not selective, whereas your cartoon is something that makes its effect by exclusions.

McLuhan: No, the TV camera is very much more selective and abstract than the picture camera. In fact, what makes the pictorial as opposed to the iconic is the inclusion of a vast amount of detail. The filling-in of space with great quantities of detail is what the Renaissance painters learned to do in the third dimension. They discovered that you could take space and fill it with objects. The iconic man doesn't fill space with objects. He makes space. The icon form tends not to be three-dimensional space so much as a made, shaped, molded, modulated resonating space.

Kermode: This would be true, I think, of a cartoon, truer than of a television picture.

McLuhan: You think of the TV picture as much filled-in with detail.

Kermode: Well, it's not selective, that's to say if the mike drops down into the picture and the camera records it.

McLuhan: Compared to a photograph of similar size, it is enormously selective and abstract.

Kermode: Well, the next point, the camera is working with its dots and out of these dots we by our collaboration make a picture.

McLuhan: There's much to fill in.

Kermode: There's a good deal to fill in, but this is something that we've got a lot of practice at doing. It's a natural, physiological movement by which we do it, isn't it? This doesn't seem to me to be the same thing as the way in which we supply the missing elements of a cartoon, for example.

McLuhan: There's more to fill in.

Kermode: There's more to fill in but it's also a different kind of filling-in. What the television camera is trying to do – because it's working on material which retains images like a retina – is give us something rather like the kind of picture we're accustomed to, whereas the cartoon deliberately does not do so.

McLuhan: This is tricky terrain, Frank, because many people are disturbed when they learn that natives and preliterates cannot see pictures, that, in fact, they do not use their eyes the way a literate man has learned to use his eyes. An Arab has great difficulty in recognizing a postcard or snapshot of a camel or any other object of everyday familiarity. John Wilson of the African Institute here said that after twenty years of teaching writing by film in Africa, he decided to teach writing so they could see movies. They have to be taught to see movies the way we are taught to read and write. It is not something that people are naturally able to do. What we call the picture, the well-filled-in, naturalistically and realistically filled-in picture, is by no means visible to the non-literate man.

Kermode: But when they do learn it, they scan it, don't they? In other words, there is a sequential element, a successive element involved in both this kind of image and in the typographical image.

McLuhan: The picture has a great deal in common with the printed word, which is in effect a series of little black-and-white pictures and images which you scan in sequence. On the other hand the picture, because it gives you a vast amount of data in a single flash, tends to be very much of a gestalt, and you can't spell out all the data that you perceive in a single instant.

Kermode: But to talk about "in a flash" and the "gestalt" is really to beg the question of simultaneity, isn't it?

McLuhan: There tends to be a great deal that is simultaneous in such a form of experience.

Kermode: If you feel that we are going to have to come to terms, or we are coming to terms much more than typographical man, with this kind of instantaneous image, I'd like to ask you about the distinction that you draw between different kinds of media within the electric technology. You call some, such as television, cool, and some, such as radio, hot. What does this mean?

McLuhan: It has to do with the slang phrase "the hot and the cool," which puzzles many people. The way it's used in slang reverses the meaning of cool. "Cool" in the slang form has come to mean involved, deeply participative, deeply engaged. Everything that we had formerly meant by heated argument is now called "cool" in slang. The idea that cool has reversed its meaning has some bearing on the fact that our culture has shifted a good deal of its stress into a demand that we be more committed, more involved in the situations in which we ordinarily work.

Kermode: A cool medium is one in which the definition is low and the audience has to work and supply the gaps.

McLuhan: Yes, like the cartoon you see that we were mentioning . before. This is real cool. Jazz as compared with classical music has many of these aspects of discontinuity and very much room for fill in. But where the information or data level is low, the fill-in or participation is high. If you fill the situation with complex data, the opportunity for completion fill-in is less and participation is less.

Kermode: There's a kind of paradox, there are many paradoxes, but the one I'm thinking of at the moment here is that a lot of people would suppose that TV is something before which you slump.

McLuhan: They're paying attention only to the programming, the content, which has nothing to do with TV.

Kermode: That's right. So you would say that the fact that some people may still be struggling to follow our conversation, that's not what we mean by, it's not our conversation that's cool. Our conversation is hot, presumably, is it?

McLuhan: Well, insofar as we're managing to be relatively detached and urbane we're really square.

Kermode: We're square?

McLuhan: We're a couple of squares, all right, as far as any cool audience is concerned, but that is a very complicated matter though. It's a new syndrome, as it were, in our culture, this cool business, and it isn't easy to unravel. On his show a few weeks ago Jack Paar asked some young person, "Why do you use the word *cool* in the sense that we mean by hot?" "Well," the youngster said, "because you old folks had used up the word *hot* before we came along."

Kermode: That just means the young are behaving in a nice, Hegelian way in the sense that perhaps your galaxies are Hegelian. Or perhaps you would you dispute that?

McLuhan: Are you thinking of the systole/diastole of forms? Yes, I think there is something of that in them, all right, this whole drive toward role-playing and depth-participation. The need for commitment and involvement began, though, with our poets and artists a hundred years ago. They began to devise opaque and difficult forms that were real cool, Rimbaud and Baudelaire and onward. They began to create these forms that required that you do a great deal to complete the meaning. Remember Edgar Allan Poe's invention of the detective story at the time of the symbolist poem. He devised this form in which the reader in effect made the story by filling in all sorts of clues and gaps; it became part of the creative process. The discovery that you could really involve the reader or viewer in the creative process by leaving out a great deal began over a century ago.

Kermode: And this is presumably now on the way out, isn't it, this kind of story, this kind of demand?

McLuhan: The connected story line as a means of organizing data, yes, is dropping out even in film with Fellini or Bergman.

Kermode: It depends on a different notion of what a logical or intelligible sequence of events is. We're passing out of that into the electric which, as you repeatedly say, is an aural kind of culture. This reminds me to ask you what I think is pretty important about your work, that if you've got phases like this which are determined technologically, one cannot only speak about the state of affairs we now have, one can also to some degree do some prediction. I think in *Understanding Media* you sometimes write as if we'd got on, if we'd pushed on deeper into electric technology than we actually have. But you do venture some predictions about the kind of life, the kind of quality of feeling that we're going to have with the new alteration in our senses. Could you say something about that?

McLuhan: I remember when I was here two years ago after a long absence, I was quite startled at the upsurge of regional dialects in England as compared with twenty years earlier, and the relative decline of standard and homogeneous English, and the quite proud display of dialects that I had hardly heard before when I lived here. This drive in depth toward regional depth of culture is a normal feature of electronic forms because of this circuitry that involves us deeper and deeper in ourselves. The French separatists, for example, at the present time in Quebec are very much related to this new image they have of themselves since television – a depth image.

Kermode: And this is something you get in the typographical heartland in the nineteenth century, don't you? Matthew Arnold has a good deal of this emphasis on regionalism.

McLuhan: I was thinking, though, more of the so-called nationalist movements of the Irish, the Scots, the Welsh, and now the French Canadians and many other areas – this sudden insistence upon ancient folkways and depth in roots of culture.

Kermode: On the other hand, nationalism in itself is typographical, you say.

McLuhan: Yes, you see the term is used loosely like the hot and the cool to fit quite contrasting situations. Nationalism originated in a highly visual unity when people could for the first time see themselves as a single public or a single social group, and it came long before there was any theory of nationalism. The event occurred before the theory.

Kermode: When shall we come to live in the global village, as you call it? When shall we have lost the sense that we have now of being individuals and of being nations, which, I think you say, is going to be one of the effects of the electric media?

McLuhan: I think an amazing amount of it has dissipated right now. The sort of disturbance that is felt in the United States in recent years, the difficulty they now have in focusing their own image as quite strangely and powerfully represented by the Goldwater mood, as it is called, is an attempt to rediscover the old blueprint of culture that they started out with in the eighteenth century, to get back to the old ground rules of culture, because at present the United States is undergoing a depth movement into itself – this teenagerism being only one aspect of it. The attempt to rediscover itself in much greater depth than it was ever founded in, is disturbing the whole image of the American at the present time.

Kermode: The vision of the future that your book could leave one with is a Big Brother image in a sense. You speak of, for example, programming cultures. For instance, you say that if the South African scene looks like it is getting too hot because of an overdose of radio then we program a lot of television, cool them off, this kind of interference with what the typographical literate man calls human rights.

McLuhan: We have never stopped interfering drastically with ourselves by every technology we could latch on to. We have absolutely disrupted our lives over and over again.

Kermode: Do you think this might lead us into a kind of electric totalitarianism?

McLuhan: No, I think the logic, if unimpeded, the logic of this sort of electric world is stasis.

Kermode: Is there a terminus? Or should we always go from the thesis of typography to the antithesis of electricity?

McLuhan: If the natural play of circuitry is depth and ever-increasing involvement and responsibility, it would seem that it does demand a great increase of human autonomy and human awareness. I think if there is a logic, and a hopeful one that appears in this, it is the dispelling of all unconscious aspects of our lives altogether; that in order to live with ourselves in such depth, in such instant-feedback situations, we have to understand everything so that our easygoing lolling about in the lap of the unconscious cannot endure, that we will have to take over the total human environment as an artifact.

Kermode: So the print-made split between head and heart that you speak of in one of your books will be healed.

McLuhan: Completely. If we have used the arts at their very best as a means of heightening our awareness of the otherwise unconscious environment, then turning a whole skill to the making of the environment itself into a work of art, namely, of transcendent awareness, would seem to be the logic of this form. And the programming of environments as artifacts, as works of art, is something that people have been moving towards on many hands. Even town planners are a familiar example of this attempt to fashion the total environment as if it were artifact instead of just introducing artifacts into environments. I think up until now mankind has been content to introduce artifacts into environments that are otherwise beyond control, to use the arts as a control. The possibility of using the total environment as a work of art, as an artifact, is a quite startling and perhaps exhilarating image but it seems to be forced upon us. The need to become completely autonomous and aware of all the consequences of everything we're doing before the consequences occur is where we're heading.

Kermode: Well, Marshall, I hope the people out there have a clearer notion now of the difference between a medium and a message.

McLuhan: I hope so too, but I don't think that clarity is necessarily the thing to discover in the beachhead. We've really established a beachhead.

---

1. Silvio A. Bedini, "The Scent of Time: A Study of the Use of Fire and Incense for Time Measurement in Oriental Countries," *Transactions of the American Philosophical Society* 53 (1963), pp. 5-47.

The Medium Is
the Massage
(1966)

On May 7, 1966, McLuhan delivered a public lecture in New York City at the Kaufmann Art Gallery of the 92nd Street Y. It was the last in a series of six programs arranged by the Foundation for Contemporary Performance Arts in association with the Y Poetry Centre.

McLuhan titles his lecture "The Medium Is the Massage," a play on his most famous aphorism, "the medium is the message." Drawing a distinction between the two phrases, he sets out to explain his thesis that the medium is "the massage, not the message, that it really works us over, it really takes hold and massages the population in a savage way."

■ ■ ■ ■ ■

I have been introduced recently as Canada's revenge on the United States, you know, from the land of the DEW line, the early-warning system. This is one of my themes tonight, the artist as early-warning system for new media. Another main theme will be that the medium is the massage, not the message; that it really works us over, it really takes hold and massages the population in a savage way.

From the land of the DEW line, I can bring you a special kind of world of jokes that you may not have encountered, namely the grievance joke associated with French Canada. It's not like the Pole joke or the Italian joke. It was Steve Allen who said that the funnyman is a man with a grievance. And if you look around at the joke world from that point of view, it's sometimes rather instructive. In Canada we have jokes such as the one about the mouse being pursued by the cat, and finally hiding under the floor, and all is quiet until suddenly there's a kind of a "bow wow arf arf" sound. And the mouse figures, ah, house dog has come, scared cat away, and up comes the mouse, the cat grabs it. As the cat chews the mouse down, it says, "You know, it pays to be bilingual." There are a lot of those jokes around up there, and I think it is amusing to watch the joke as a sign of grievance.

In the same way slang can be an indicator of subtle change of sensibility, shifting patterns of perception. And if our youngsters were encouraged to pay the utmost attention and respect to slang from this point of view, it might clarify some of the problems they have. Isn't it strange that no matter how complicated slang is, no child ever made a mistake in slang, and that goes even for the word *cool*. I once presented one of my children with a strange object. It was actually a yearbook from Rice University, and it was about six by six inches square, just cards taped together, and it had the little stories and poems of the year. When our thirteen-year-old saw this, he said, "Dad, that's real cool," and he was using the term quite correctly. But many people have difficulty with that term, especially when I use it. Actually, I merely draw it from the world of slang in which it means involved and detached at the same time like a surgeon operating.

"Cool" means identification with the creative process. When a person is both involved and detached, he has to identify with the creative process. You can read a great deal about the concept of cool in Mr. [T. S.] Eliot's *Four Quartets*. It's a somewhat Oriental concept. And Western man tends to be either involved or detached, but not both at once. So this is a new development in our society. It is a complicated idea, and it is a profoundly creative idea.

One of the strange changes that is taking place in our world – John Cage recently gave me a curious example of what it meant to him – is that we're moving out of the world of the planter, out of the world of the specialist, out of the world of the fragmented person into the world of the hunter, the unified person. This is the meaning of James Bond, it's the meaning of the sleuth, the meaning of the obsession with crime in our world, because the criminal and the criminologist alike are both hunters. They belong to the old paleolithic world and are really a new type of human being in our midst – the person who explores the total human environment the way the old hunter and the food-gatherer used to explore his entire environment as a unified field. Man the planter, man the basket-weaver, and man the pot-maker came in after the paleolithic man, and we've had thousands of years of the planter. When I mentioned this to John Cage, he said, "You know, that's very interesting. I spend my life hunting mushrooms. I am not the least bit interested in cultivating them." This is a curious illustration of the difference between the two kinds of man. The hunter is not concerned with classification or specialism or the processes of cultivation, only with discovery.

So there are some rather exciting things afoot. For example, if one wanted to talk about the future of work in our world, we know that it has very little to do with jobs as they're presently understood. It has much to do with discovery, much to do with knowing, much to do with involvement in processes so much so that the future of work is plainly knowing.

There was a time when the job of royalty, of the prince, was not to learn any special skill, but simply to grow up. Jacques Ellul, the author of *Propaganda*, points out that in our twentieth-century world the child is the hardest working child in all human history. Our children in a twentieth-century information environment have to process more data than any human being in any previous culture of the world. Our children from early infancy are engaged in extraordinarily hard work, and that work is mainly just growing, growing up because to grow up in a modern electronic environment is a fantastically complex and difficult job. It's also a job which threatens to deprive people of identity, the personal concept. One of the peculiarities of an electronic environment is that people become so profoundly involved in each other that they lose that sense of private identity. This is one of the peculiar cruxes of our time that people, precisely because they become profoundly involved in one another in an all-at-once simultaneous field of happenings, then begin to lose their sense of private identity because identity used to be connected with simple classification and fragmentation and non-involvement. In a world of profound involvement, identity seems to evaporate.

The future of work as knowing, the future of identity as involvement, brings one around to the idea of perhaps roles instead of jobs. A mother doesn't have a job; she has sixty jobs, and that's a role. A top exec doesn't have a job; he has many jobs simultaneously, and that is a role. I think this is happening to all of us in varying degrees. The very speeded-up world we live in guarantees that we can all have a highly integrated existence which really points in the direction of role-playing rather than of job-holding.

I was talking to a group of managers recently, and I thought I'd try it out on them. It seemed to me plausible to suggest to them that the real future of old age in business is discovery. They were quite struck by this. It seemed to make profound sense, most of them being over forty. But the point about old age as a period of discovery contradicts the merely chronological approach to things. The older man in business is a man who knows the field, and he's in a much better position to be a discoverer than some younger man who is merely job-holding. And so there's hope.

There's another basic theme associated with this computer threat to job holding. "Come into my parlor," said the computer to the specialist. Anybody who wants to be taken over by a computer should just specialize. But one of the hopeful things about the computer is this: that as a retrieval system, the instantaneous speed of retrieval by computer offers a tremendous future of discovery, because a retrieval system of very high speed brushes so many facets of knowledge together, so many kinds of layers of experience get brushed together that they reveal structures, they reveal forms, they reveal the life of forms, they reveal knowledge in all sorts of new patterns. So it is typical of our time, or any other time, perhaps, that they should tackle this new form of technology and put it on the old job-classified data. What we've been doing with the computer up until now is merely giving it to the librarians and to the card cataloguers. In other words, let the new form do the old work.

But the possibility of retrieval as a form of discovery is something we all know from our own memories. When we recall something, we tend to make discoveries. We certainly tend to make changes in the pattern of the thing we know. *Finnegans Wake* is a book built entirely on this retrieval system: "Casting her perils before swains."[1] This is a phrase out of *Finnegan*, and it's typical of the technique of *Finnegan*. Joyce regarded the human language as the biggest storage system of human knowledge and perception anywhere. And his own use of that vast store of human perception was to retrieve from it in a pattern which made discovery natural: "Though he might have been humble there's no police like Holmes." And that's retrieval. The pun as a form of retrieval brushes various layers and contexts together and creates a new discovery pattern. So the computer offers many advantages besides just the threat to wipe out jobs and provide leisure.

Under electric conditions, "the future of" suggests an approach to various other things which I might mention while on that theme, the future of the planet as a work of art. Since the planet has as its new environment satellites and electronic information, it is rapidly becoming an old nose cone, a work of art like the Model T. Like old silents on TV, the planet is now becoming the content of our own man-made environment. And the future of the planet as Williamsburg, as work of art, as carefully prepared archeological exhibit, is really quite amusing. And the future work of mankind will be mainly raking and tidying and clipping the old planet, putting it in shape, reconstructing it pretty much the way it was when the pilgrims landed.

I'm merely using this as an illustration of a basic theme that anything becomes a work of art as soon as it is surrounded by a new environment. Give any form at all a new environment and it acquires art status. In other words, it becomes visible because this is the peculiar character of art. It creates attention; it creates perception. And the role of the artist as a creator of perceptual models and perceptual means is perhaps misunderstood by those who think of art as primarily a blood bank of stored human values. In our time there are a great many unhappy people who see the great art treasures of the past being polluted by a corrupt new vulgar environment. It never occurs to them that perhaps their job is the penetration and exploration of these new environments, and that the mere accumulation of past human experience in blood banks called art does not really contribute very much toward the perception of our current environment.

A great many people are confused and unhappy at the present time because they sense that all the great values of the past are being polluted and corrupted and anemicized by vulgar entertainment, mass entertainment. In that regard let's pause a moment on the future of the book. The printed book created what we call the public. With the coming of electric circuitry, you have what is called the mass. It's a time factor. The book created the public because the book, as a printed form, made possible a very large body of people but not simultaneous. The peculiar dimension of the mass audience is that it's all-at-once. It's a happening like telegraph, like the daily paper. The daily paper is a good example of a happening because everything that's in it happens at the same moment. The dateline is the only organizing principle in a newspaper. There is no connection between the items otherwise than the dateline. And if you take away the dateline from any newspaper whatever, you have a quite handsome surrealist poem. It becomes much more enjoyable if you pull the dateline off. The dateline presents a plausible pretext of rationality, meaning, connection, which is not really very deeply related to the newspaper.

The sort of constellation of events that is incorporated on any page of a newspaper is a kind of environmental image for which many people find it natural to identify empathetically, and so on. And the paper as a happening is very much in this dimension of the all-at-once. The editorial page by contrast represents a point of view, a fixed, arrested position with regard to the happening. The editorial is very closely related to the book as a form.

The printed book as a form created the public, and Montaigne said upon recognizing the existence of this new entity, this new environment called the public, "I owe a complete portrait of myself to the public."[2] It was a matter of immediate response on the part of writers that their task as writers would be from now on the etiolation, the evolving of self-portrait, an image. There was no such impulse or response in the Middle Ages, when they had the scribe-and-manuscript culture, or in the ancient world. It was with the coming of the printed book that people suddenly felt the need to reflect, to bounce their image off this public as a form of self-expression, self-portraiture. Self-portraiture was a very special response to the printed word and was never felt as a need or a response for the manuscript form of culture.

Xerox has brought a kind of revolution into the publishing world that is only being felt slowly. It will be felt more and more. Xerox is the application of electric circuitry to a world which had formerly been merely mechanical and fragmented. Xerox or xerography enables the reader to become publisher, and this is an important aspect of all electric circuitry. The audience is increasingly involved in the process. With print, the audience was detached, observant, but not involved. With circuitry, the reader, the audience becomes involved in itself and in the process of publishing.

The future of the book is very much in the order of book as information service. Instead of the book as a fixed package of repeatable and uniform character suited to the market with pricing, the book is increasingly taking on the character of service, an information service, and the book as an information service is tailor-made and custom-built.

The strange dynamic or pattern of electric information is to involve the audience increasingly as part of the workforce instead of just tossing things to it as consumer or as entertainment. The tendency is to involve the audience as workforce as with electric toasters, electric razors, electric anything. You do the work. And we're coming to this. I'll come back to this perhaps apropos the future of television or forms like that in which it will become possible or is now possible to brief audiences in prime-time on top-level problems in physics, science, whatever, and invite their response to these problems by IBM cards sent out in supermarkets or in magazines. Robert Oppenheimer is fond of saying that there are kids playing right here on the sidewalk who could solve some of my toughest problems in physics; they have modes of perception that I lost forty years ago. Oppenheimer realizes in that remark that most scientific problems are really not concept problems but percept problems, that most scientists are blocked in their perceptions by preconceptions and prepossessions.

When you're dealing simultaneously with several million people, it's obvious that somebody in that audience is going to have a perceptual perforation into the problem without any difficulty whatever. Eight scientists working on a problem for fifty years might not get through, but ten million people working on the problem for ten minutes might get through.

Audience participation can mean many things. You remember in the old quiz shows the audience was browned off when it discovered that the shows had been rigged, and they had really been left out. There was nothing dishonest about rigging quiz shows any more than there is about rigging a movie and making it look different from what a real-life situation might. But it was a misuse of the medium of TV, which calls for much participation. The movie world doesn't call for any participation. It's a fantasy world, highly visual with the audience sitting very much back from the show. But TV is not like that. TV is a profoundly involving medium with the audience as environment, the audience as vanishing point, the audience as screen.

In the movie, the audience is the camera; the audience looks out at the environment. With TV, the audience is the environment, is the screen, is the vanishing point. And this creates a completely different relation to programming. It creates a kind of Oriental effect with reverse perspective. Audience as vanishing point creates a reverse perspective with a kind of Orientalism involved. It's not accidental that our Western world is taking on many profound Oriental traits since television. But it's not just television. It's electric circuitry.

The safety car is another peculiar example of electric circuitry afflicting a mechanical object, that is, the consequences of the car are now being felt by the designers of the car whereas formerly they merely concerned themselves with the projectile. You hurled it out into the public and let the chips fall where they may. But suddenly in the age of circuitry the feedback is coming into the design. Product design now takes on the character of audience participation.

But perhaps the safety car is more an example of the effect of the space capsule. Bucky Fuller is famous for having pointed out, among many brilliant observations, that the space capsule is the first man-made environment. Totally man-made. You know the old saying, "You can't take it with you." With the space capsule you've gotta take it with you. You have to take the planet with you. So the space capsule is really a kind of complete environment in which all consequences of all maneuvers are anticipated by design, by the built-in design pattern.

This sense of the space capsule as applied to the motor car is now creating the motor car as a padded cell. And what more appropriate form of design for the maniac driver? But the motor car as a feedback design pattern turns into a padded cell. I have a friend who speaks of the motor car as the "carsophagus." "Don't drop the bandwagon" is another of his remarks. The carsophagus is a form of peripatetic vehicle in which the effects of the object on the environment are built back into the object. This is something that is characteristic of electric circuits. The circuit as such is a form that feeds back and feeds us into the circuit. All circuits are do-it-yourself objects.

The book on Xerox becomes somewhat like the safety car. The reader becomes publisher and author. In the ancient days of the scribe in manuscript cultures, the scribe was both publisher and author. And we are sort of circling around to that condition again in which with Xerox the reader of the book can excerpt it and relate it to any other excerpts or any other notions or any other soundtracks and publish it as his own creative effort. And this is already happening. Teachers are strongly inclined to make their own books for their own classes designed specifically for the needs of a particular class. This is another peculiar feature of electric technology; it tends toward decentralized, tailor-made, custom-built servicing.

It's typical of our rear-view mirror orientation that we look at all these new technologies as if they were reflexes of the old technology. For heaven knows how long, people have encountered each new technology by translating it back into the old familiar one. You know all the types of examples of that. The first motor cars were made with buggy whip holders. And the new automation computer devices are being made as if they were card catalogues. This strange habit of looking backward when moving forward is one that perhaps won't quite bear up, isn't quite consistent with jet speeds and electronic speeds. It may be that we are the first human generation to feel the need for a close look at what's actually happening under our noses instead of that rear-view mirror look. The rear-view mirror is a very comforting spot. It gives you a sense of remoteness like *Bonanza*. The suburban world of our time lives in *Bonanza*-land. This gives it a comfortable sense of distance and security. It's a nineteenth-century image. When the railway was new in the nineteenth century, people formed an image not of the railway and the new world it was creating, the new cities. They formed an image of an Arcadian retreat, a pastoral, innocent world, a Jeffersonian paradise. That was their image of the railway. As soon as they banged into the railway, they translated themselves back into an agrarian, Arcadian pastoral world. This is a normal human reaction to novelty and innovation. Its inappropriateness as a reaction in no way deters people from this strategy. It's a kind of déjà vu. Every time you encounter the new, you say, "I've been here before, and this isn't as new as you might think."

This seems to tie in with the ancient theory of human learning as needfully going from the known to the unknown. I don't think that there's too much evidence that this is a basic pattern of human learning. I think there's lots of evidence, though, for the fact that the moment you encounter the unknown you translate it back into the known. This means that we never encounter the unknown. We encounter only convenient self-deceptions. The world of electric technology and circuitry as involving the audience as workforce has extraordinary implications. You can see while this development is taking place, while the audience is becoming profoundly involved in making a world of entertainment, the gloom voyeurs stand around classifying the program and providing program ratings of the most unhappy kind. While the audience is about to take over this vast creative act, the gloom voyeurs stand around reading off audience ratings. For example, 350 million people each week watch *Bonanza* in sixty-two different countries of the world. What could be a more gloomy image of the human condition? Actually, *Bonanza* isn't that bad a show. There was a little boy in Nairobi a few weeks ago, in Africa, who was found wandering around the city, and on being questioned by the police, he explained he had come from some village, and he was an orphan. And they said, "Yeah, well why are you here?" He said, "I'm looking for Mr. Cartwright. I think he can help me." And I'm quite sure he could, too.

For countries like Africa or Indonesia or China, *Bonanza* must have a very strange aspect. It must look like some distant science-fiction image of some remote world to come, that is, for countries that never had a nineteenth century. And by the way, this also happens to apply to the West Coast. California didn't have a nineteenth century. During the nineteenth century people were making their way out there as best they could. There were no factories, there were no cities, there was no industrial life. The result is that, not having had a nineteenth century, the West Coast leapfrogged out of the eighteenth into the twentieth century with great advantage to itself. That is, if you leapfrog out of one century and leapfrog over a century or two, you can, upon encountering the new conditions, feel completely uncrippled by the older ones. You can behave in a much more spontaneous and much more resourceful way than people who have been through the whole process.

Those who lived through the nineteenth century lived through the most stupefying and fragmenting and systematizing and categorizing century in human history. Therefore, anybody who leapfrogged over the nineteenth century might retain some imaginative life. And those who went through the nineteenth century were maimed and mutilated very horribly because of this enormous power to fragment and systematize and classify everything that happens.

Canada is in a peculiar position. Canada never had anything but a nineteenth century, that is, Canada was invented in the nineteenth century. The United States was invented in the eighteenth century. Wasn't it Adlai Stevenson who said, "Columbus went too far"? This business of leapfrogging is not without its significant features. In our present time there are many countries in the world who are leapfrogging out of 10,000 B.C. into the twentieth century. Many countries are doing this, leapfrogging out of prehistory into post-history. Just psychically, what is the consequence of skipping thousands and thousands of years of Western history? No one knows. It hasn't been thought about.

And so to come back a moment to the strange prodigy of the xerographed book, the tendency to involve audience as actor, as publisher, as writer is a characteristic that applies to an enormous amount of electric technology quite apart from this particular instance. It's a sort of prospect that will increasingly bring us into the need for new instruments of handling our problems. One way of putting it is to say, for the first time in human history, there is more information and data outside the classroom or the school situation than inside. The sheer amount of levels of information outside in the environment far exceed the amount of data and information inside the classroom. This is not just of very recent origin. It's occurring more and more rapidly and on a much bigger and bigger scale. You know all the statistics one hears, again intended to panic people, that by the year 1980, there will be more scientists than people, and more and more. What would seem to be, therefore, the future of education in a world in which the proportions of information have been reversed? In ordinary human past, there'd been more in knowledge and information available inside the classroom conditions than outside. With this spectacular reversal of this condition, it would seem to be possible that the business of the school has also reversed, that the business of the school is no longer instruction but discovery. And the business of the teaching establishment is to train perception upon the outer environment instead of merely stenciling information upon the brain pans of children inside the environment.

We have never had an educational system programmed to train perception of the outer world, and it will create a considerable trauma or shock to switch into that mode of activity. But it's like the audience in the entertainment world that is ready to go into action on high-level problems. The children in our school system at the present time have very high-level information resources both in their own makeup and nervous systems and in the world immediately accessible to them, and they can be given the task, in teams, of researching all sorts of fascinating problems. For example, suppose a group of elementary schoolchildren were given the job of trying to find out what was happening to America since color TV. Teams of them could be turned loose in just making inventories of effects observable in clothing, in food, in cars, in designs, in family life, in all sorts of forms. If they merely made inventories, they didn't have to use any ideas, they didn't have to reach any conclusions, they merely were invited to observe, to make inventories of effects. Small children love this kind of work, and are just as able researchers as any team of sociologists in the world. And the accumulation of such inventories can be absolutely necessary and indispensable to the highest operations.

I was hearing a sociologist today suggest that it was quite impossible to make a direct connection between television and any change whatever. This is in a certain sense true. On the other hand, if you make an inventory of all the changes that have taken place in the last twelve, fifteen years since TV, without any preconceptions, without any theories about what television is or what television does, if one merely made an inventory of all the changes in speech, in politics, in habits of human association, habits of reading, habits of dress and food, it might become quite evident that there had been a kind of a pattern of change. And this would be a scientific report in the usual way of an induction. No conclusions. Just here are the changes that have taken place in the past fifteen years in all these fields since TV. And you could in turn look at that same curve in relation to some other technology if you chose.

The way to study the effects, for example, if you wanted to study what the motor car was, you might find out more from what it did to the environment and the community. In other words, if you notice that the motor car created road systems and vast servicing industries, you would probably have a better idea of the motor car than you could have from photographic images of cars in catalogues. The car is what it does to people and what it does to the environment, and so with any other medium, as far as I know. A medium creates an environment.

An environment is a process; it is not a wrapper. It's an action, and it goes to work on our nervous systems and on our sensory lives, completely altering them.

The content of any technology is inevitably the older technology. The new environment goes around the old environment, and turns the old one into an art form. This happened to the Greeks, it happened to the Romans, it happened to the Middle Ages, it happened to the Renaissance, and so on. There's no lack of evidence of this happening. At the present time, I mentioned the strange fate of the planet as art form resulting from the new environment of satellite and electronic information.

One of the biggest revolutions in Western culture occurred when the world of the arts in the eighteenth century began to deliberately create landscapes and environments as a way of controlling mental life. Whether in painting or in poetry or in landscaping and city building, the eighteenth-century artist created landscapes, what he called the picturesque, making things look as if they already were framed in pictures. And he found that these landscapes or picturesque situations could be used as a means of controlling the mental life of the observer. The function of Romantic art or picturesque art was to control moods. It was all oriented toward effect. Earlier the arts had been more concerned with simply permitting the audience to enjoy the sense of community and festive involvement. Medieval arts, as we are popularly aware, were festive and communal and participative. They were not intended to give any sense of privileged or elite life.

It's a little bit like the Balinese who say, "We have no art. We do everything as well as possible." They think of art as that which applies to the environment. They program the environment. Art for them means dealing with the environment itself as if it were a work of art. This is one of the meanings of pop art. Pop art is a recognition that the outer environment is itself capable of being processed like art. And this, I think, is an electronic phenomenon that only in an age of electronic immediacy and totality could we ever dream of tackling the whole human environment as a work of art. This is happening to us. And pop art is merely a report to the nation that this event is taking place and that we can prepare ourselves accordingly.

You know how the elite artists, the blood-bank people, are shocked by pop art because they try to classify the content of pop art. Instead of viewing pop art as a means of perception, they think of it as a means of classification, as merely a means of ratings. The idea that it might have a function just training human perception doesn't seem to get through easily to these people.

But let's go back for a moment to the eighteenth-century landscape picturesque artists. Having devised this landscape as a way of controlling the mental life of populations, there came shortly after, in the middle of the nineteenth century, a sudden break in which the artist said it is the inner life, not the outer landscape, that counts. And the symbolists – the Rimbauds, the Wagners, the Baudelaires – began to examine the inner mental landscape as something that could be programmed as a means of enriching consciousness, of creating an inclusive consciousness. Instead of just creating a mood as the Romantic artists had done, they began to work on the idea of the total consciousness. And they became intensely interested in the mythic consciousness, in the corporate consciousness of the mythic man, and in the Jungian archetype. This was a big moment in the history of art and culture that concerns us because they discovered that the meaning then of the work of art was not as a conveyor or a package, but the meaning of this was as a probe, an exploratory probe into the outer world.

You will find this quite unanimously expressed by Flaubert and Baudelaire and others. Flaubert's way of putting it was simply, "Style is by itself an absolute way of perceiving things."[3] That style is not a way of expressing something. It's a way of seeing, of knowing. I derived all my knowledge of media from people like Flaubert and Rimbaud and Baudelaire. They began to study the materials with which they worked in order to be faithful to style. They deliberately began to study the matters, the materials with which they worked. You find Frank Lloyd Wright and modern architects doing this at the turn of the century. They began to study not what they wanted to express, but what were the means available for expression. When they began to study these materials, they quickly discovered that the medium is the massage or message. This was a big breakthrough, because what they discovered was that the function of art is to teach human perception. Conrad's remark that's often quoted, "It is, before all, to make you see!"[4] – not that you may know this or that, but that you may have the means of perceiving.

This idea of training all our resources upon a complex world which needs perception very badly, this idea that the future of the future is the present, this idea that if you examine the present deeply enough you will find all possible futures, this became a very common idea to the symbolists and to people around the later nineteenth century. It's an idea that gives art a profound role as a teacher of perception rather than as a conveyor of some precious content. I know there's a lot of controversial territory in between those two positions. This strange clash between the old Romantic art of the outer landscape and the new symbolist art of the inner landscape or of human consciousness, inclusive human consciousness, was already a kind of indicator of the strange condition in which we are today, in which we begin to extend the very means of awareness into the environment, in other words, extending consciousness.

There are some lessons to be learned from the artist in regard to our own time in this way. A man like Seurat, the painter, was painting TV in 1880, that is, he had a model, a perception that had all the technical features of TV – rear projection, minute fragmentation of image into these little pointillisms. The TV viewer receives these little spots on himself, they wrap around him, he becomes lord of the flies. They settle on him literally. The TV viewer is covered with these little dots, these little flies. Seurat knew that, and he painted that back in 1880, but he wanted an effect. He was not painting the technology. He was painting the effect he wanted. The effect he wanted from this TV image or this type of pointillism was that of total audience participation, involvement of a multi-sensuous kind. Seurat was consciously aiming at not merely visual but total sensuous involvement. He wanted all the senses to be involved in the painting, and that's why he used this rear projection. There's a pleasant story about Malraux taking General de Gaulle around an art gallery recently and pointing out this and that. And de Gaulle was saying, "Uh, ah, what's that one over there?" and Malraux said, "That's a Dufy." "And what's that one?" "Oh, that's a Renoir." "Ah," says de Gaulle suddenly, "I know that cartoon over there, that must be a Rouault?" "No, sir, that's a mirror." Rouault came to mind because he too was a rear projectionist, that is, the light came through the image to the viewer. This light through or rear projection creates a much higher degree of involvement than otherwise is possible. The painters suddenly wanted to involve the audience deeply in their images with all their senses.

Electric technology, both radio and TV, for example, have this power of involving us in all our senses. Many people don't seem to recognize very easily that a cartoon, a mere outline, a mere contour is a profoundly involving experience that requires participation of all our senses. Whereas a photograph is more limited to one sense, merely the visual, a cartoon, although it seems much simpler, is actually more profound in involving our senses, our sensuous lives.

The painters knew this. I learned this from the later-nineteenth-century painters, all this stuff about the media that I talk about I learned from them. In fact, you'll find the great instructors in all media matters are these painters and poets of the later nineteenth century and people like James Joyce and Eliot and Pound and others. They spent their lives studying our senses as they go out technologically into the environment because they realized this had a profound effect on language and on the medium that they were working with as poets.

One of the things that puzzles many people about our world of advertising is that there seems to be such a strange disproportion between the effect of the ad and the content. Take a Picasso painting such as *Man Sitting in Chair*. You see no chair. You see no man. But the painting is designed to convey the effect of sitting in a chair, not the appearance. Not what it looks like, but what it feels like. This painting the effect minus the appearances is a means of involvement, and the painters and poets alike sought increasingly the techniques of involvement in our time. Hence, the return, for example, of poetic drama, because the dramatists discovered that if you want to really involve the audience in the action of a play you must not use prose, which merely leaves them talking at each other and to each other, the audience, and the actors. You must use poetry which brings the audience right up onto the stage.

It's rather paradoxical to say that poetry is a more involving experience than prose when in actual fact most people read prose and very few read poetry. But in terms of effect this is true. And it's the same with painting. So now in modern advertising there's much the same revolution going forward as has been going forward in painting and poetry and in the world of entertainment with the audience moving more and more into the act. Modern advertising is more and more a substitute for the product. Because of its rich resources and means of sensuous solicitation, modern advertising can give you the effect of almost anything without your having to bother about having the thing at all.

Tony Schwartz is in the audience tonight, bless him, and he is an expert in this very department. In terms of auditory magic, the power of giving the effect without the thing is entirely within his domain. Advertisers have been puzzled for a long time in their surveys to discover that people don't really read ads until they own the thing. If you own that icebox or that car, then you notice the ads very much because this is how you get your satisfaction. The ad world then becomes an enormous area of information service, part of the service industries, enabling you to substitute the effect for the thing, just like Picasso. And this is called non-objective abstract art. Now you see it, now you don't see it. The artists usually go through these maneuvers fifty or more years before the engineers get around to them. And so the advantage of knowing art is orientation in a complex world. It gives you a fifty-year breathing space before the thing hits, and that is a considerable advantage.

One of the peculiar results of this profound involvement through circuitry as compared with the old detachment of literacy and visual culture has been the increasing Orientalization of the Western world. Even as we pour and lavish our nineteenth-century technology on the East, we are doing to ourselves the exact opposite. We are Orientalizing ourselves by going inward. It's very easy to see them going westward, but very difficult to see ourselves going inward because it is so environmental with us that it becomes almost imperceptible. You can, however, if you do inventory, discover the amazing rise of Oriental values in the arts and in all sorts of produce and all sorts of habits. And I'm sure they could by similar inventories find that the West made considerable inroads into their world.

This may serve as an example of the kind of change that is going on in our own midst, Orientalizing ourselves while Westernizing the East, and seemingly taking much credit for the Westernizing and not even noticing what we're doing to ourselves. We are doing this to ourselves; nobody is doing it to us.

Many people find it desirable to saddle me with moral judgments and the need for value judgments. If you notice this Orientalizing of the Western world, it would be impertinent and somewhat fatuous to approve or disapprove of anything that concerned the life of nations and many, many millions of people simultaneously. It would be very much like the position of Margaret Fuller, one of Bucky's relatives, by the way, who said back, what, in 1850: "I accept the universe." And Emerson, I think it was, who said, "She'd better!"

1. James Joyce, *Finnegans Wake* (London: Faber & Faber, 1939), p. 202.

2. Michel Eyquem de Montaigne (tr. Donald M. Frame), *The Complete Works of Montaigne* (Stanford, Calif.: Stanford University Press, 1957), pp. 677-78.

3. Gustave Flaubert (tr. Francis Steegmuller), *The Selected Letters of Gustave Flaubert* (New York: Farrar, Straus, & Cudahy, 1953), vol. 2, p. 358.

4. Joseph Conrad, *The Nigger of the "Narcissus"* (London: J. M. Dent & Sons, 1950), p. x.

Predicting Communication
via the Internet
(1966)

By by the mid-1960s, McLuhan had made the world aware that television was a medium that held modern man in its thrall in profound ways that did not meet the eye, and he did this in the most old-fashioned way possible by saying it to as many people as he could.

On May 8, 1966, This Hour Has Seven Days, a Canadian Broadcasting Corporation TV public affairs program, featured McLuhan in an interview with the journalist Robert Fulford. McLuhan has the novel idea that the teenager of the mid-sixties was a far more realistic, serious, and meditative creature than the teenager of the previous generation, all because of television and its "involving" quality.

In this interview McLuhan also accurately predicts the sort of "interactive" communication that has become possible in the past decade via the Internet.

McLuhan: The planet is going to get a great new processing from the meteorologists and from all sorts of scientific therapists. It's going to be put in apple-pie order so it will be nice to come home to once in a while, back to the old homestead from outer space every once in a while.

Fulford: You've been writing about the mass media for a good many years and now you're an object of the mass media. How has this changed your view of it, if at all?

McLuhan: Let me instead explain why this has happened, because, if you notice, the mood of North America has suddenly changed very drastically. Things like the safety car couldn't have happened ten years ago.

Fulford: Why is that?

McLuhan: It's because people have suddenly become obsessed with the consequences of things. They used to be obsessed with mere products and packages and launching these things out into markets and into the public. Now they've suddenly become concerned about what happens when these things go out onto the highway, what happens when this kind of program gets on the air. They want safety air, safety cigarettes, safety cars, and safety programming. This need for safety is a sudden awareness that things have effects. Now my writing has for years been concerned with the effects of things, not their impact, but their consequences after impact. Unlike the fantasy world, the escape world of movies, TV creates the enormously serious and realistic-minded sort of person, well, almost Oriental in his inward meditativeness.

Fulford: This is the teenager of today?

McLuhan: Yes, he's becoming almost Oriental in his inwardness.

Fulford: He's so thoughtful and serious.

McLuhan: Yes, grim, whereas the movie generations of the twenties and thirties were a coon-coated bunch of superficial types, had a good time and went to college but not for knowledge and that sort of thing. All has changed.

Fulford: And changed because of television?

McLuhan: Very much. Television gave the old electric circuitry that was already here, gave it a huge extra push in this direction of involvement and inwardness. You see, the circuit doesn't simply push things out for inspection. It pushes you into the circuit. It involves you. When you put a new medium into play in a given population, all their sensory life shifts a bit, sometimes shifts a lot. This changes their outlook, their attitudes, changes their feelings about studies, about school, about politics. Since TV, Canadian and British and American politics have cooled off almost to the point of rigor mortis. Our politics require much more hotting up than the TV medium will give them. TV is ideal when you get two experts like ourselves discussing TV. This is good TV because there's a process going on of mutual challenge, discovery, and processing. Now TV is good for that, and the same with ads. If the audience can become involved in the actual process of making the ad, then it's happy. It's like the old quiz shows. They were great TV because it gave the audience a role, something to do. They were horrified when they discovered they'd really been left out all the time because the shows were rigged. This was a horrible misunderstanding of TV on the part of the programmers.

In the same way, most advertisers do not understand the TV medium. Do you know that most people read ads about things they already own? They don't read things to buy them, but to feel reassured that they have already bought the right thing. In other words, they get huge information satisfaction from ads, far more than they do from the product itself. Where advertising is heading is quite simply into a world where the ad will become a substitute for the product, and all the satisfactions will be derived informationally from the ad, and the product will be merely a number in some file somewhere.

Instead of going out and buying a packaged book of which there have been five thousand copies printed, you will go to the telephone, describe your interests, your needs, your problems, and say you're working on a history of Egyptian arithmetic. You know a bit of Sanskrit, you're qualified in German, and you're a good mathematician, and they say it will be right over. And they at once xerox, with the help of computers from the libraries of the world, all the latest material just for you personally, not as something to be put out on a bookshelf. They send you the package as a direct personal service. This is where we're heading under electronic information conditions. Products increasingly are becoming services.

Fulford: What kind of a world would you rather live in? Is there a period in the past or a possible period in the future you'd rather be in?

McLuhan: No, I'd rather be in any period at all as long as people are going to leave it alone for a while.

Fulford: But they're not going to, are they?

McLuhan: No, and so the only alternative is to understand everything that's going on, and then neutralize it as much as possible, turn off as many buttons as you can, and frustrate them as much as you can. I am resolutely opposed to all innovation, all change, but I am determined to understand what's happening because I don't choose just to sit and let the juggernaut roll over me. Many people seem to think that if you talk about something recent, you're in favor of it. The exact opposite is true in my case. Anything I talk about is almost certainly to be something I'm resolutely against, and it seems to me the best way of opposing it is to understand it, and then you know where to turn off the button.

The Marfleet Lectures

(1967)

The Marfleet lectureship was endowed in 1910 by Lydia Marfleet in memory of her husband, Pearson Kirkman Marfleet, an Illinois businessman with an abiding interest in the relationship between Canada and the United States. The first Marfleet Lecture was given by President Howard Taft (1857–1930), who spoke on the presidency of the United States. The second was given seven years later by Sir Robert Borden (1854–1937) on constitutional development in Canada. The political forum was subsequently abandoned in favor of lectures by leading scholars and intellectuals from across North America.

On March 16 and 17, 1967, McLuhan delivered a two-part Marfleet Lectureship in Convocation Hall at the University of Toronto. A week before his scheduled appearance, Newsweek *ran a cover story (March 6, 1967) titled "The Message of Marshall McLuhan," which made the lectures immensely popular with the general public as well.*

The title of the two-part lecture is "Canada in the Electronic Age." The theme of the first, "Canada, the Borderline Case," is the strange effect of being on so many borderlines in Canada. The second, "Towards an Inclusive Consciousness," is the retribalizing effects of electronic media.

In the first lecture, McLuhan describes what it means for Canada – culturally, geographically, and historically – to be on the border with the United States. "Nature and history seem to have agreed to designate us in Canada for a corporate, artistic role. As the U.S.A. becomes a world environment through its resources, technology, and enterprises, Canada takes on the function of making that world environment perceptible to those who occupy it." In McLuhan's view, Canada's role in the world is that of a counter-environment to the United States. He imagines Canada as a frontier country, "the land of the DEW line which is part of our invisible environment. It is a frontier of pure information typical of a variety of frontiers . . . that have emerged in the twentieth century, and which have altered our entire relation to ourselves and to our world."

In his second Marfleet Lecture, McLuhan explains that we have suddenly been thrust into a world where everything happens at once, where the same information is available at the same moment from every part of the world, and where the past is instantly retrievable. This all-at-onceness creates a kind of total memory which is a return to tribalism in the sense of a comprehensive inclusive consciousness. McLuhan likens the situation to that of the Inuit fifty years ago confronting the railway: "We today in the electronic age are all as primitive or bewildered as any Eskimo ever was when he saw his first railway train. Our ability to cope with this new electronic technology is not any greater than that."

CANADA, THE BORDERLINE CASE

I asked a friend who lives in Windsor to provide me with a few borderline jokes from the Customs and Immigration Division, and it was a fairly happy thought. The first one he came up with concerned a guard, an immigration officer, who put his head in the window of the car and said, "Where were you born?" And the voice said, "Toronto." And he said, "Get over to the side!" Anybody who doesn't say "Trona" is obviously a fake. Toronto has been a wonderful place to live for the last two decades, and my family appreciates it as well as I. To mention another of these borderline stories before dropping them. After questioning a man and woman in a car coming from Detroit one December afternoon, an officer asked a five- or six-year-old boy standing in the back seat, "Whose little boy are you?" He didn't answer, and so the man said, "Who do you like best?" He replied, "I like daddy best cause he's got my Christmas present hidden under the front seat."

There are quite a lot of these borderline stories if you could bear them. There are simply endless stories of this type. In the days when the ballpoint pen was first introduced, it cost as much as seventeen dollars in Canada, much less in Detroit. One night a local man of the cloth was stopped at the tunnel and questioned about twelve pens he had in his breast pocket. He protested vigorously, but when he was finally convinced that he should pay the duty and go forth and sin no more, he raised his coat for his wallet and exposed a carton of Lucky Strikes in each hip pocket.

I have worked on this subject, thought about it for weeks, and I find it's very difficult to think about a Canadian audience as an entity, as a coherent, unified thing. Was it Jean Marchand the other day who said that as a result of his touring of Canada he decided that Canada is five countries between British Columbia and the Maritimes, five countries. It's very difficult to address five countries simultaneously, and I think this is perhaps one of our strengths also.

Nature and history seem to have agreed to designate us in Canada for a corporate, artistic role. As the U.S.A. becomes a world environment through its resources, technology, and enterprises, Canada takes on the function of making that world environment perceptible to those who occupy it. Any environment tends to be imperceptible to its users and occupants except to the degree that counter-environments are created by the artist.

A *New Yorker* cartoon a few months ago showed two fish that had climbed out on the shore. One said to the other, "This is where the action is." A wit has said we don't know who discovered water, but we're pretty sure it wasn't a fish.

The one thing you can never see is the element in which you move. Canada, of course, is the land of the DEW line, which is part of our invisible environment. It is a frontier of pure information typical of a variety of frontiers, of which I will have some explanations later, that have emerged in the twentieth century, and which have altered our entire relation to ourselves and to our world. For the value of a frontier as a sort of interface or complex process of continuing change adds greatly to the powers of human perception and growth. For example, a writer seeking to discern the contours of the Victorian age hit upon this striking approach.

The chief turn of nineteenth-century England was taken about the time a footman at Holland House opened a door and announced: "Mr. Macaulay." Macaulay's literary popularity was representative. It was deserved, but his presence among the great Whig families marks an epoch. He was the son of one of the first friends of the Negro whose honest industry and philanthropy were darkened by a religion of sombre smugness, which almost makes one fancy they love the Negro for his colour and would have turned away from red or yellow men as needlessly gaudy. But his wit and his politics combined with that dropping of the puritan tendency, but retention of the puritan tone which marked his class and generation, lifted him into a sphere which was utterly opposite to that from which he came. This Whig world, the great new industrial tycoon world, was exclusive but it was not narrow. It was very difficult for an outsider to get into it, but if he did get into it, he was in a much freer atmosphere than any other in England. Of those aristocrats, the old guard of the eighteenth century, many denied God, many defended Bonaparte, nearly all sneered at the Royal Family. Nor did wealth or birth make any barriers for those once within this singular Whig world.

The value of such a frontier between worlds consists in enriching all of them by a kind of process of dialogue and interaction that would be quite impossible within any one of them – the great value of Ireland.

Well, in that connection the famous episode in Macaulay's early life concerned a letter he received from [Francis] Jeffrey, the editor of the *Edinburgh Review*, who said, "Mr. Macaulay, the more I think, the less I can conceive where you picked up that style."[1] He got it on the frontier between these two families. It's an interplay between styles. The great value of Ireland in English life appears sufficiently in the work of Bernard Shaw. His play, *John Bull's Other Island*, is a dramatization of the frontiersmanship that Shaw exploited with poetic imagination:

When I say that I am an Irishman I mean that I was born in Ireland, and that my native language is the English of Swift and not the unspeakable jargon of the mid-XIX-century London newspapers. My extraction is the extraction of most Englishmen: that is, I have no trace in me of the commercially imported North Spanish strain which passes for aboriginal Irish: I am a genuine typical Irishman of the Danish, Norman, Cromwellian, and (of course) Scotch invasions. I am violently and arrogantly Protestant by family tradition; but let no English Government therefore counter my allegiance: I am English enough to be an inveterate Republican and Home Ruler. It is true that one of my grandfathers was an Orangeman; but then his sister was an abbess; and his uncle, I am proud to say, was hanged as a rebel. When I look around me on the hybrid cosmopolitans, slum poisoned or square pampered, who call themselves Englishmen today, and see them bullied by the Irish Protestant garrison as no Bengalee now lets himself be bullied by an Englishman; when I see the Irishmen everywhere standing clearheaded, sane, hardily callous to the boyish sentimentalities, susceptibilities, and credulities that make the Englishman the dupe of every charlatan and the idolater of every numbskull, I perceive that Ireland is the only spot on earth which still produces the ideal Englishman of history.[2]

This may happen to Canada. After all, with Raymond Massey we have already produced the ideal typical American of all time, Abe Lincoln. He concludes:

Personally I like Englishmen much better than Irishmen (no doubt because they make more of me) just as many Englishmen like Frenchmen better than Englishmen . . . But I never think of an Englishman as my countryman. I should as soon think of applying that term to a German. And the Englishman has the same feeling. When a Frenchman fails to make the distinction, we both feel a certain disparagement involved in the misapprehension.

Macaulay, seeing that the Irish had in Swift an author worth stealing, tried to annex him by contending that he must be classed as an Englishman because he was not an aboriginal Celt. He might as well have refused the name of Briton to Addison because he did not stain himself blue and attach scythes to the poles of his sedan chair. In spite of all such trifling with facts, the actual distinction between the idolatrous Englishman and the fact-facing Irishman, of the same extraction though they be, remains to explode those two hollowest of fictions, the Irish and the English "races."[3]

Perhaps an even more striking instance of the operation of frontiersmanship in intensifying human perception and creativity occurs in the career of James Joyce, of whom a fellow Irishman observed not long ago that the oddity of James Joyce seems partly that of a prodigious birth out of time, another frontier, an oddity favored, certainly, but not engendered by the artistic climate of the twentieth century. Ireland, owing to her isolation from the European development, also in part no doubt to foreign domination, had produced no important body of literature during the Middle Ages, an age which in her case has continued almost to the present day.

Joyce is Ireland's first great native writer, her Dante or her Chaucer. Though expressing his age as every writer should, it was also necessary for him to express in his manner those varied ages to achieve a great collective Yeatsian dreaming-back. He took with immense seriousness his destiny of forging the uncreated conscience of his race, something which is given only to the artist to do, so that he had to be by turns a St. Augustine crying aloud his sins, a scholastic glossing on

Aquinas should Auld Aquinas be forgot, the producer himself of a summa or great synthesis, and finally a Duns Scotus of splitting hairs and mangling words. And all the time he was essentially a humorous, skeptical Dublin observer and Everyman among artists with a schoolboy love of puns.

When *Ulysses* appeared in 1922, T. S. Eliot wrote in his review:

> In using the myth, in manipulating a continuous parallel between contemporaneity and antiquity, Mr. Joyce is pursuing a method which others must pursue after him [that is the parallel between modern Dublin and ancient Ithaca, Homer's world]. They will not be imitators, any more than the scientist who uses the discoveries of an Einstein in pursuing his own, independent, further investigations. It is simply a way of controlling, of ordering, of giving a shape and a significance to the immense panorama of futility and anarchy which is contemporary history.[4]

Eliot is specifically saying here that the only way you can control any very complex anarchic situation is by parallel borders, a border or a frontier. "It is a method already adumbrated by Mr. Yeats, and of the need for which I believe Mr. Yeats to have been the first contemporary to be conscious."[5] It is also Mr. Eliot's method, since he is a man from Missouri who set up his cabin on the banks of the Thames in 1915 approximately, an example of premature brain drain. There won't be time to go into that theme, but for anybody who has studied frontiers in literature, they'll know the drain is never in one direction only. Eliot went to England and made visible for the first time to the twentieth-century English the new world of French literary and artistic discovery. The environment of European art of the past decades had become so invisible to the English that it took this American to reveal it by his artistic activities.

Yeats himself was a great frontiersman, and there's a famous passage in Yeats in which he says that one of the defects of French literature is its lack of double plots or parallel structures, without which you cannot achieve the emotion of multitude. It is by a parallel between two kinds of actions which Shakespeare frequently uses that you achieve the emotion of multitude. This is a state in which we live constantly, that is, on the border. We live constantly in two worlds, and we have endless access to this resource, which enables us to create an emotion of multitude in various patterns.

The world of Whitman doesn't need very much propping up in order to be seen as a frontier world. His barbaric yop, which he insisted on sounding across the roofs of the world, was very much born on the frontier and directed to the frontier. He was also the author of that famous phrase "Passage to India." He has a poem of that title. And this is another kind of frontier famous in the work of E. M. Forster. *A Passage to India*, a novel whose main action is a parallel between East and West, is by far his greatest work, and it derives its greatness from this amazing parallel of actions that are seemingly totally unconnected. There doesn't have to be any connection between the actions as long as they continue parallel to one another. What Columbus failed to discover, namely, a passage to India, these artists did work at. Was it here that Adlai Stevenson said, "On this platform Columbus went too far"?

The sense of the creative possibilities of the frontiers, personal or national, is as prominent in the work of Yeats as in anybody. A famous observation of Yeats is that the borders of our minds are ever shifting, and that many minds can flow into one another and create or reveal a single mind, a single energy, also that the borders of our memories are shifting, and that our memories are a part of one great memory, the memory of nature herself. This great mind and great memory can be evoked by symbols – by artists, in other words – and I think that Canada has contributed some amazing frontiersmen, of whom surely Northrop Frye is as extraordinary as any with his frontiersmanship between the world of literature and the unconscious. This has given him a world position, and it is very much a frontier activity.

One of the most breathtaking frontiersman of culture was Edmund Burke, whose eighteenth-century view of the Revolutionary War in America anticipated Harold Innis's conclusion to his book *The Fur Trade in Canada*. Innis argued that the war was basically a conflict between the fur traders and the settlers.

> The fur trade interests in the East naturally regarded settlement as a threat to the trapping lines and the fur supplies. The East has always feared the result of an unregulated advance of the frontier and tried to check and guide it. The English authorities would have checked settlement at the headwaters of the Atlantic and allowed the savages to enjoy their deserts in quiet lest the peltry trade should decrease. This behaviour called out Burke's protest that the Atlantic seaboard entrepreneurs were prepared to call off the development of America in the name of the fur trade.

Burke protested:

> If you stopped your grants, what would be the consequence? The people would occupy without grants. They have already so occupied in many places. You cannot station garrisons in every part of these deserts. If you drive the people from one place, they will carry on their annual tillage, and remove with their flocks and herds to another. Many of the people in the back settlements are already little attached to particular situations. Already they have

topped the Appalachian Mountains. From thence they behold before them an immense plain, one vast, rich, level meadow – a square of five-hundred miles.

He was not entirely correctly informed.

Over this they would wander without a possibility of restraint; they would change their manners with the habits of their life, would soon forget a government by which they were disowned, would become hordes of English Tarters, and, pouring down upon your unfortified frontiers a fierce and irresistible cavalry, become masters of your governors and your counsellors, your collectors and controllers.[6]

Burke saw centuries ahead, as frontiersmen often do. A frontiersman of a very different kind was James Boswell, who invaded the British metropolis to obtain the scalp of the great panjandrum of English letters. Boswell and Johnson never ceased to dialogue back and forth across the frontier. Boswell would say: "But Scotland, sir, has many noble wild prospects," and Johnson: "But, sir, let me tell you, the noblest prospect which a Scotchman ever sees is the high road that leads him to England."[7]

The Scot in England is transformed as radically as Shaw claims for the Irishman in England. Surrounded by staid and unimaginative humanity, the Scot in England becomes daring, vigorous, and efficient. In Canada, in a more inspiring setting, he is inclined to lethargy and caution lest he upset so ideal a milieu. In Scotland, the Scot is quite another person.

I had a delightful conversation one day with John Wilson of the African Institute of London University. Speaking of frontiers, he's a Scot who spent much of his life in Africa trying to teach natives to read by means of film. He gave up after twenty years and taught them to read their letters so that they could see the film. He discovered they couldn't see films until they'd learned to read and write. But John Wilson said that it's true that the frontier turns the Scot into an artist, an entrepreneur. But at home, Wilson pointed out to me that he is a very unimaginative bureaucrat. The Scot at home is not the Scot who created the frontiers of the world.

Frontiers have many patterns that are often difficult to discern. The frontier in American history became visible as a social and geographic factor only with the advent of the telegraph, which eliminated geography. I am grateful to our Alexander Lecturer, Professor Frank Kermode, who pointed out that in 1867 the telephone was invented but also *Das Kapital* was published. As a frontier event that's rather memorable; something for Expo '67 to play up, maybe. With the coming of the telegraph, the old environment went inside and became much more discernible when it had this new surround. If the railway created a kind of unity in Canada, a kind of continuity in space, the CBC as a new frontier in space and time tends to eliminate the geographic pattern of the railway in favor of a very much more massive, and at the same time more discontinuous, pattern of electric information. We have yet to discern the features of electric environments. They are so little like any of the earlier environments in the world that we have yet to learn really how to cope with them at all.

Harold Innis became keenly aware of the oral tradition in the ancient world as a result of his studies in the modern world. That is, modern anthropologists dealing as they do with non-literate societies have made our contemporaries specially aware of the integrating and unifying force of oral culture. With this awareness has come our knowledge of the disintegrating power of civilization. Whereas Rousseau in the highly civilized society of the eighteenth century had come to value

the integral qualities of primitive life, we have come to understand in detail the operation of those factors that enable
civilized values to permeate oral situations.

It is much less feasible to observe the factors now operating
upon ourselves, factors that are creating a new oral culture
overlayering our inherited civilization. A civilization is always
based upon some written form. Professor W. T. Easterbrook,
who is spending the year in Tanzania, is very much aware of
these matters, and he often drops notes portraying the amazing
differences between these oral and semi-literate people and
ourselves. He says, for example, "Jokes as such are very rare
over here but the element of fun is everywhere, delight in the
quick thrust, the parry and return. This breaks through at
every meeting no matter how serious the subject under consideration. There is no bitterness or malice in the delightful play
of wits, no resentment when a thrust strikes home. There is a
lightness of touch that most of us have lost, and I do hope that
with development it will not be lost here."

The qualities of an oral society are becoming more evident
to us in the electronic age with the tremendous increase of professional wits and entertainers. The Bob Hopes are basically
oral types, not to mention the Jimmy Durantes. But it is much
less feasible to observe the factors now operating upon ourselves, factors that create a new oral culture. We are all familiar
with the great flood of sounds that environ our world, but
we're not necessarily familiar with the fact that electronic
simultaneity and all-at-onceness is itself auditory in structure
even when there is nothing to be heard. That is, auditory space
or auditory effects are from all directions at once as in this
auditorium, whereas the visual world is not from all directions
at once. The civilized world deals with one thing at a time.

Our new electric power enables us to put the human
unconscious outside as an unintentional environment that we
experience unconsciously. A new bridge or frontier is thus

created between the conscious and the unconscious. One of the anarchic, and to any adult one of the utterly confusing things about our time, is this sudden appearance of a world in which everything that man ever was or ever knew is outside as part of a human environment. This is very much like the unconscious, and to confront the unconscious as an environmental fact is to seem to confront or encounter anarchy. Our teenagers respond to this unconscious with delight as a kind of Disneyland of imaginative effects. Our teenagers have been called the last generation, meaning not lost nor last in the sense of never any more, but last in the sense of ultimately summing up all other generations.

The deep interest of Harold Innis in the interplay across the frontier of the written and oral forms of human experience may well have inspired his old friend Eric Havelock to make a splendid and unique study of the oral tradition in ancient Greece. Eric Havelock's *Preface to Plato* is a study of the social frontiers between the written and the oral traditions and their effects on the shaping of human perception and behavior in the ancient world. Quoting him for a moment, "To repeat then: in an oral culture the hoarded usages of society tend also to assume the guise of hoarded techniques."[8] The poets in that sense become the tribal encyclopedias of the time of the culture, and the tribal encyclopedia does not just contain thoughts or observations but also techniques about how to conduct society, how to govern it, how to control it. Hence, he argues, Plato's hostility to the poets as educators, he being a new type of educator. This is Havelock's thesis in that book, that the poets were the first educators of Greece. They were the educational establishment bitterly resented by the new revolutionaries coming in with the written word, the Platos and others. Plato's famous war on the poets is simply a war of one educator against the educational establishment. But that's only incidental to the theme of the book. The theme of the book is

that an oral society packs, records in its oral encyclopedia, techniques for operating a society, and I think you'll find that the
teenagers today tend to take that outlook on even such things as
the hit parade. They seem to regard it as absolutely indispensable for survival. It's operational. It isn't entertainment at all. It's
part of their way of life. It is a needed and central fact. Our
teenagers are all behaving like members of an ancient oral
culture, surrounded as they are by rather recently civilized
people. There's a huge confusion between these two conditions.

"The most striking example," he said, "as furnished in the
first book of the *Iliad* is that of the practices of seamanship, a
craft central to Greek civilisation at all periods. The poet's
narrative is so composed that opportunity is afforded for a sea
voyage. The girl, if she is to be restored to her father's shrine,
must be transported on shipboard. This becomes the occasion
for recapitulating some standardised operational procedures,
which are spelled out in four distinct passages forming a pro-
gressive pattern, as follows." I'll only read the last one.

> Let us advisedly gather and thereupon a hecatomb
> Let us set and upon the deck Chryseis of fair cheeks
> Let us embark. And one man as captain, a man of counsel,
> there must be.[9]

In our own world, the fragments of the oral tradition
mainly exist in religious ritual and liturgy, which have taken on
a new significance and relevance in our electronic time. In fact,
we are in a sense playing backwards this process described in
Preface to Plato. We are moving from the written to the oral at
a much higher speed than the Greeks ever disintegrated their
oral culture by means of the written word.

It is a great shock to many to discover that the demands of
oral culture are deep and involving compared to the patterns
of written and literary culture. When our children encounter

the old literary establishment of classified knowledge, they feel rejected and ejected by superficiality. Incongruously, we classify them as dropouts.

Introducing the mass media, the Massey Report records:

> Before proceeding to the problems of broadcasting, of moving pictures and of the other "mass media" in Canada, we think it worthwhile to point out that about one half of the Canadian population was born earlier than 1923 and that most of these older members of our population spent their formative years in a society where radio was unknown, where the moving picture was an exceptional curiosity rather than a national habit, and where as a consequence the cultural life of most communities centered about the church, the school, the local library and the local newspaper.
>
> It is probably true, for example, that Canadians now in their thirties or older will recall that the church organist and church choir provided much of the music of their earlier years. More often than not the organist in English-speaking Canada was from the old country, trained in the English tradition of organ and choral music. . . . The great musical events of the year were usually the concerts given by the local church choirs, aided by a visiting celebrity. Although the radio has vastly increased the size of listening audiences, we must not forget that long before its day there flourished in towns and cities of Canada a vigorous musical life, or that the musical tastes of a considerable part of our population were in large measure formed by the well-trained musicians who came to us, bringing with them a tradition of fine music.[10]

This is, in effect, a statement that traditional music offers a kind of oral encyclopedia, a tribal encyclopedia, which was basically a form of mass media that the Massey Report has found more acceptable than the new mass media. But it's an interesting sidelight or insight to say that the tradition of church music really was a kind of corporate, communal art shared by all, and that bound the community together.

The principal theme of *The Double Hook* by Sheila Watson is the effect of people in a simple frontier community in British Columbia trying to create a sort of unity in their inner lives by forming images of social cohesion and communication. "His mind sifted ritual phrases. Some half forgotten. You're welcome. Put your horse in. Pull up. *Ave Maria. Benedictus fructus ventris. Introibo.*"[11]

What all of us do, only the artist makes visible. The ordinary procedures and environmental patterns of a society don't become visible until the artist creates this counter-environment of art objects. This is a frontier problem that relates to Canada's position as a frontier country, giving Canada a kind of world art role in making visible the vast, man-made American environment that is becoming a world environment.

I don't know whether that makes much sense at first, but as the United States becomes a world environment, some means is needed to make it visible and capable of appraisal, appreciation, and criticism – that, only the artist can do – and Canada is essentially in this artistic role. It isn't the sort of thing that can be casually talked about on a large public occasion but it needs chatting about and sifting and thinking.

Art makes the corporate and communal accessible to the individual, whose task it is to assimilate the tradition, to modify it in relation to the new situations that are continuously forming. The artist does this by creating counter-environments as mirrors of the present. Emily Carr's paintings made accessible to us all the British Columbia experience for the first time for many of us. The Group of Seven made Canada visible.

There's a very peculiar example of the Canadian role in this matter in the States. There is a great oral tradition related to the name and figure of Paul Bunyan and his blue ox, Babe. This great French-Canadian oral tradition swept through the American frontier at the turn of the last century and is not really much known to Canadians, but it is very vividly known to Americans, at least in the Midwest. I encountered the Paul Bunyan stories and adventures and hyperboles and fantasies first at Madison, Wisconsin, where they were very prevalent. Wisconsin was a great logging territory and Paul Bunyan was a French-Canadian logger, and all these wild stories concerned these superhuman feats that he and his ox performed in the logging communities.

The oral tradition, this huge epic oral tradition of Paul Bunyan, seems to be a natural result of the frontier meeting of two worlds that tends to create this emotion of multitude and bigness which I mentioned earlier. And perhaps the most striking example of all is Ned Pratt or E. J. Pratt's poetry. His epics, his "Cachalots," his epics about the great effort at creating the CPR – *Towards the Last Spike, The Titanic, Brébeuf and His Brethren* – all these epics are very much frontier poems, the meeting of worlds, the artistic effort to make visible two worlds at once, frontier worlds. Ned Pratt is essentially an oral poet, as anybody could tell from meeting him and hearing his tales and anecdotes, and he's our English equivalent of the French-Canadian Paul Bunyan, I mean our English and literate equivalent of the Canadian oral Paul Bunyan.

The artist is by nature a boundary-hopper and a claim-jumper. Let me suggest that Canada as a counter-environment to the world environment created by the U.S. may not only be the means of creating a colossal artistic vision of the present and the meaning of the U.S. in the world, but that Canada may well incur the deep unpopularity that results from performing this necessary function. The artist as a sort of frontier boundary-hopper has often been regarded as a kind of enemy of society because he makes things, he points out things that many people would rather not notice. One way to grasp the importance of this artistic vision can be shown by relating our situation to comparable patterns of culture in other times.

That is why I'd like to take a moment to relate our situation today to those crucial developments in ancient Greece which led to the founding of the Western tradition. In a basic sense we are retracing in the present age many of the cultural stages that men traversed at that time, thus making that period of the world very visible to us. This retracing is in its turn manifested in another pattern of events closely related to our history, in which we see the United States performing in Asia the same frontier push against the tribal man that we associate with Europe and the beginnings of American history. This is altogether so obvious that apparently it is not ordinarily verbalized.

What is now going on in Vietnam is very much a repeat of what went on in the American frontier for centuries. And it may be merely a sort of automatic gesture of unconscious frontiersmen going through the old motions of being frontiersmen without noticing what they're doing. The Passage to India has taken a rather drastic turn. No, but I mean once you've just studied the pattern, quite apart from assessing it as a policy, once you begin to look at that pattern of the frontier and its continuous, incongruous new milieu of the East itself, this is altogether startling.

To be on the frontier, then, between worlds, living in divided and distinguished worlds – as Sir Thomas Browne said – the human condition, whether geographically or historically, is to have a keener vision of present events than is available to those directly involved in them. The name Keenleyside was given, I am told, to Scots who dwelt by Hadrian's Wall, a frontier. They appear to have been accredited with a more-than-ordinary degree of awareness.

Fifth-century Athens was the first human community to experience the shift from oral to visual in literate culture. It was the first human community to detribalize and to disintegrate its oral tradition. It was the first to discover the privileges and the anxieties of individual identity and, as you are familiar, our time is obsessed with this problem of identity and the quest for identity. But for the Greeks the discovery of individual identity was terrifying and altogether shattering as it was dramatized, for example, in *Oedipus Rex*. It wasn't a very welcome experience to discover what Oedipus discovered – that all men were bound so intimately to one another as to be in effect living in a perpetual incest. This was one of the immediate discoveries of civilized man when he faced his tribal heritage.

The present century is the first in which detribalized Westerners have begun to experience the reverse shift from written to oral culture. Today we experience that kind of bewilderment that ensues when an individual and private culture begins to resume an involvement in the corporate and collective modes of awareness with all the depth and commitment that implies. Our most immediate experience of this takeover of individual culture by the corporate and collective vision of tribal man is in our own homes. The present generation gap between teenagers and their parents is a major manifestation of a technological gap between the mechanical and the electric cultures.

While we exert ourselves to confer Western technology and patterns on the Oriental world and backward countries everywhere, we are less likely to notice that the inner trip of Oriental culture has grown apace in our own society, thanks to the operation of electric technology on our sensory order. As we Westernize the Orient and Africa, we have been even more successful in Easternizing and tribalizing ourselves. Westernizing is a process that we can easily perceive in the rear-view mirror. We naturally feel at home with that process after 2,500 years of cultivating it. The tribalizing process, the inner trip, the depth involvement, and the experience of a unified human family, this is something of which we have had no experience for many centuries. It is a process that is located so entirely in the present that it does not appear at all in the rear-view mirror which we habitually look to for reassurance and nostalgic orientation.

I'm going to stop right there because this brings up my theme for tomorrow. Having talked tonight about a divided country or culture, tomorrow we will take the time to discuss a unified one.

1. G. Otto Trevelyan, *The Life and Letters of Lord Macaulay* (New York: Harper & Brothers, 1878), vol. 1, p. 117.

2. George Bernard Shaw, *John Bull's Other Island* (London: Constable & Company, 1931; revised for Standard Edition, 1947). Preface to Politicians: What Is an Irishman? pp. 15-16.

3. Ibid., Preface to Politicians: Our Temperaments Contrasted, pp. 17-18.

4. T. S. Eliot, "Ulysses, Order, and Myth," *The Dial* 75 (November 1923), p. 483.

5. Ibid.

6. Edmund Burke, *On the American Revolution: Selected Speeches and Letters*, ed. Elliott R. Barkan (New York: Harper & Row, 1966), p. 90.

7. James Boswell, *Life of Samuel Johnson*, ed. G. B. Hill and L. F. Powell (Oxford: Clarendon Press, 1934), vol. 1, p. 425.

8. Eric Havelock, *Preface to Plato* (Cambridge, Mass.: Harvard University Press, 1963), p. 81.

9. Ibid.

10. Massey Report, Royal Commission on National Development in the Arts, Letters, and Sciences (Ottawa: Printer to the King, 1951), p. 19.

11. Sheila Watson, *The Double Hook* (Toronto: McClelland & Stewart, 1959), p. 45.

■ ■ ■ ■ ■

TOWARDS AN INCLUSIVE CONSCIOUSNESS

This is an extraordinary day in the history of culture. This is St. Patrick's Day and in Irish Scottish territory such as part of my family comes from one hardly knows how to tackle the occasion. But there is one kind of story that seems to serve to straddle all the facets of this occasion. And it's a true story, concocted by Father Sheridan at St. Mike's who has quite a brogue himself, and part of it has slipped away from me. He did a round, a sort of set of observations on the different colleges, pointing out the appropriateness of their traffic arrangements in terms of their religious affiliation. And he started at UC, and that's the bit that now eludes me. But when he got to Trinity, he said, "Now, there you have a very significant thing – a cop – just what is appropriate to a church with a big tie-in with the establish-ment. And you go down the way to Vic, and there you have a push button, just what is needed for private interpretation. And now you get down to St. Mike's, and there is no push button, no cop, but we have extreme unction."

Last night I talked on the theme of a divided consciousness, a frontier land, with its special opportunities to enlarge and deepen images. Speaking of the DEW line and radar, which is one of the great components of our world, albeit invisible, one of the peculiar structural features of radar is what is called interference. Images are recorded two ways, from the front and from a kind of flow of energy that is called interference. By means of this interference, an image that might otherwise merely be a flat picture can be seen in all dimensions in depth, in multi-dimensions, and radar uses this kind of depth hologram as well as the straight head-on impact. The double border effect has some of that hologram possibility of seeing things in more than one aspect simultaneously, a sort of multiple perspectives. The DEW line is the very essence of this kind of thing.

This kind of doubleness I had noticed as belonging in a peculiar way to the baroque period of our Western world, the great doubleness of the period of Bach and Milton and the baroque architecture of Bernini. Francis Lord Bacon said: "I have taken all knowledge to be my province."[1] He lived in this double world, and it seemed natural to him to take all knowledge for his province in a way that characterized many other people in his time, encyclopedic time. What Bacon did was to take the Book of Nature, which had been the medieval image of the natural world, and to this he added the Book of Scripture, the Sacred Page. He took both these pages and directed to these pages a kind of analytic gaze of comprehensive inclusiveness. I'm suggesting that the very components that make for a divided consciousness also can, with a certain encouragement, become the means of an inclusive consciousness such as Bacon took for granted in his own case.

Plato and Aristotle, the representatives of the new literate culture of Greece in philosophy, had this same doubleness. They straddled the written and oral traditions. They translated the tribal encyclopedia of the preceding culture into the written, classified form, and they too produced an encyclopedia, an encyclopedic philosophy. They sought, however, by classification and visual stress to purify the dialect of the tribe, to purify the tribal encyclopedia. Codifying it in written form gave them certain advantages.

To come for a moment into our own present world, a friend has said that the future of the future is the present. If you really are curious about the future, just study the present, because what we ordinarily see in any present is really what appears in the rear-view mirror. What we ordinarily think of as present is really the past. Modern suburbia lives in *Bonanza*-land. It looks back nostalgically and sentimentally to the frontier as a kind of safe, and at the same time, admirable and desirable world.

This habit of seeing back one stage when thinking that one is looking at the present is an age-old human habit, and it may be that we are the first to discover a means of overcoming the limitations of this habit. They are mostly involuntary. For example, in the Renaissance the image they saw very vividly in their rear-view mirror was a medieval one. What the Middle Ages saw in their rear-view mirror, what they thought was the present, was ancient Rome. What the industrial nineteenth century saw in the rear-view mirror was the Renaissance. And what we see in the rear-view mirror is the nineteenth century. We don't see the present, not very much of it. But we can recomfort ourselves with the observation of Norman O. Brown that every breakdown is a breakthrough. And when perception breaks down, we sometimes make discoveries that might otherwise elude us.

It's only too tempting to think of our present teenager problem, or the one that's constantly talked about, as something that had happened many times before. The rear-view mirror tells us that children have always taken this outlook. It's normal. They've always been ready to take over the world at the drop of a doubt. But I'm not sure that that is the real picture we see in front of us.

In the present age, suddenly leapfrogging out of the nineteenth century into the twentieth century, we have left behind the neolithic age with its specialisms and its stationary agriculture. What seems to have happened is that after many centuries, in fact, after many thousands of years, we have left this neolithic time and have leapfrogged into a period of the hunter once more; only we live in an age of information, and the hunter now seeks information. The hunting is now done in the form of research.

That aspect of leapfrogging out of one age into another might have some relevance if we point out, for example, that whereas Canada has a nineteenth-century origin, British Columbia never had a nineteenth century. It leapfrogged over the nineteenth century into the twentieth, and this gives the British Columbia people a very different outlook on the twentieth century from ourselves.

California never had a nineteenth century. It never had an age of heavy industry and big metropolitan areas until this century. This is in many ways a great advantage. French Canada never had a nineteenth century, never had an eighteenth century. This is not a criticism at all. It means that they leapfrogged out of the seventeenth century into the twentieth century, bypassing the eighteenth and nineteenth, and this offers certain advantages. But all over the world today there are countries leapfrogging out of remote Stone Age pasts into the present.

Our own situation today is not unlike that of an Eskimo fifty years ago confronting railway trains and things like that. We today in the electric age are as primitive or bewildered as any Eskimo ever was when he saw his first railway train. Our ability to cope with this new electronic technology is not any greater than that. We have done a bit of leapfrogging, too. But many parts of the world are leapfrogging out of areas, out of periods in time which go way back beyond the Western world altogether. What happens to them as a result of skipping our whole Western development and suddenly participating in twentieth-century information is something that needs lots of imaginative understanding on our part. I suggest perhaps that in our own case, our reluctance to abandon the nineteenth century is not entirely grievous or unhealthy. If we can hang on to this nineteenth-century beachhead a little longer, we may be able to leapfrog over the twentieth century altogether.

Now the United States is an eighteenth-century country, not a nineteenth-century country. It was born in the eighteenth century. And its outlook on a nineteenth-century country like ourselves is naturally rather confused. But it is making desperate efforts to live in the twentieth century, as we are not. They feel a certain obligation to conquer the twentieth century or take full advantage of it, and we don't feel that compulsion. It's one of those things as far as we're concerned. Take it or leave it.

But the nineteenth century was an intensely centralized social organization. Whether it was the family or the city or the church, it was all very central in its structure. The establishments of all kinds were centralized. In this time of electric information, such centralism is quite impossible. Yet we feel terribly upset by this pulling-apart of our old centralist functions, family and so on.

When the railway train first came into our land, its first effect was to take people off the land and to break up families. It was a tremendous disrupter of family life. Anybody could get on the train and go off to the big town. Most of the folklore and literature of the past sixty or seventy years is concerned with just that – families breaking up, people leaving for the big town.

The work of Veblen, a very brilliant and dazzling economist, the author of *The Theory of the Leisure Class*, is very much the expression of a country boy from Scandinavia feeling great dislike and discomfort in the presence of this new disruptive centralist kind of society that was forming around Chicago. He began to call it all sorts of naughty names – a society dedicated to conspicuous waste, conspicuous consumption, and so on. These terms have hung on. They stayed around long enough for Mr. Eliot to incorporate it in *The Waste Land*, which is an image of this new nineteenth-century city built on the principle of waste.

The end of the neolithic age is, then, the leapfrogging out of that ancient time when man the hunter settled down and began to cultivate the land and to weave baskets and make pots and tools. The neolithic man after the age of the hunter has suddenly dropped aside or dropped back, and we are back once more in the age of the hunter, only this time the hunter is a fact-finder, and a researcher, and a discoverer, a sort of James Bond/CIA type.

Now this end of the neolithic going along with the electronic is a return to the tribal encyclopedia, is a return to tribalism in the sense of a kind of comprehensive inclusive consciousness. But it's not preliterate; it's post-literate, and it's very different to be post-literate from being preliterate.

We have no precedent for this, no way of getting our bearings, nobody to tell us what's happening. We're suddenly projected into a world where everything happens at once, that is, electrically. The same information is available at the same moment from every part of the world. Energy is available everywhere at once. And electric retrieval systems enable us to recall anything at all instantly. This kind of total memory such as, for example, permits even now by micro-card the putting of all the books in the world on one desktop. Electrically, it is possible to put every book in the world and every page of every book in the world on one desktop. It's a kind of development which seems to make science fiction seem very silly. Science fiction is very far behind what is happening.

But this all-at-onceness changes our entire outlook. Notice the change in our joke styles. The new popular jokes are full of puns. You'll remember all the elephant jokes and all the Alexander Graham Kowalskis, the first telephone Pole, that type of joke. The kids went crazy about them for a while. Now they seem to have eased up. But the joke, instead of having a story line, has become a conundrum, has become a kind of puzzle. "What did the chair say to the ice box?" "Why did the moron take a ladder to church?" And all those questions. These are conundrums which are intended to test the agility. Instead of giving you a nice story with a nice conclusion, they insist that you get into the act. There's the story about the little boy who was helping a nun across the street, and she thanked him when they got across, and he said: "That's all right, Madam. Any relative of Batman is a friend of mine."

Now this kind of joke is based on a pun but it's not a verbal pun. It's a visual pun, not an auditory pun. The sort of contoured presence and ritual costume of Batman has a certain punning resonance in relation to the nun's costume. And this kind of joke is a fairly familiar type of joke. The kind of observation that Joyce never tired of is somewhat similar. It's a visual pun. It's also an auditory pun. "The urb, it orbs."[2] Under electric pressure the city becomes a globe. "The urb, it orbs" and the little round schoolhouse takes over. The entire planet becomes a little round schoolhouse. The whole environment becomes a teaching machine.

This is a kind of flip from the world of Rousseau. Rousseau in his highly civilized century dreamed nostalgically, rear-view-mirror style, of the noble savage and the unspoiled, untouched human environment, thinking of that unspoiled environment as a teaching machine that could impart perfect lessons to the inner man.

Wordsworth is full of these observations about the perfection of the natural scene as a teaching machine:

One impulse from a vernal wood
May teach you more of man
Of moral evil and of good
Than all the sages can ("The Tables Turned").

But this kind of new environment that we have, an information environment, electrically programmed, turns the entire planet into a teaching machine, and it's a man-made teaching machine. The planet is now a man-made environment. This kind of flip or reversal is the sort of thing we're living in the midst of. The old tribal man had a consciousness that was corporate. Now we're beginning to get one that is private for the first time in human history. The planet as a man-made teaching machine offers us an inclusive consciousness that is at the same time private and tribal – something in that direction – and it's a kind of development that is charged with further power of enlargement and significance. One of the results of the man-made environment of information becoming a teaching machine is that the audience becomes workforce. Instead of being there as a passive consumer, the audience has increasingly become workforce. It used to be simply in the form of opinion polls. But even opinion polls are sort of beginning to move toward the use of the audience as workforce. Even the voting machine and the polling booth represent an attempt to mobilize audiences as workforce, as creative energy in governing.

Harvey Wheeler, out on the West Coast at the Center for the Study of Democratic Institutions, has a view of democracy in which he argues that the great discovery of democracy was to transform every citizen into a guerilla fighter. The targets of his guerilla activities were the establishment. Napoleon earlier had made the discovery that every citizen can be a soldier. And once he made that discovery, his armies became absolutely irresistible. He then proceeded to use his armies as an educational force to impart the meaning of this democratic revolution to all the backward countries of Europe. Harvey Wheeler goes on to say that in a very backward area like Vietnam, for example, if every man is prepared to die, it is possible to overcome the discrepancy between the advanced American-type technology and their backward technology. The technological gap can be bridged by the willingness of everybody to die. He says this discovery had earlier been made in democracy, that if every citizen is a guerilla fighter, it is possible to educate everybody in the very act of being governed, and it is in the very act of government that the total society becomes an educational operation. The American attitude towards this duty of the citizen to attack the establishment in and out of season, to keep his eye firmly fixed on every center of power as necessarily corrupt, is quite alien to our British tradition. The British have a habit of noblesse oblige which assumes that those who govern have a kind of uprightness and integrity which entitles them to a certain allegiance. The democratic principle is more fragmentary and more skeptical and assumes that anything that looks like power is almost certainly corrupt. The job of the citizen is to dig it out, search it out, and to wreck that institution as soon as possible.

You can see a certain big step away from that attitude to power as corrupt. It was a popular observation of the nineteenth century that all power corrupts and absolute power corrupts absolutely – Lord Acton – and it was popularly received. Today, with the Peace Corps and the CYC [Company of Young Canadians] sort of enterprises, we begin to see the possibility of educating in action in backward territories or in poverty-stricken territories by simply putting young people in there in order to learn the language, the ways, the habits, to participate totally in the way of life of these underprivileged setups, using this as an educational operation, not just as a social welfare operation. It is a means of imparting the language and culture of these territories to young people from other countries. This is a sort of military operation in a quite new mode, a more benign mode, but it shouldn't completely conceal from us the possibility that military action is itself a form of education. We're teaching many hard lessons to people in Vietnam now and with vast cost to ourselves. An unlimited educational budget has been provided for this purpose. Napoleon did this. Julius Caesar and Alexander the Great were some of the great educators in the early Western world who trampled up and down over backward territories learning them, and when they had learned those lessons, they hurried back and destroyed Rome.

The kind of participation possible today for young people in other countries as a sort of complex educational venture, a kind of civilized dialogue between cultures, is a tremendous new venture on the part of mankind. The Peace Corps enterprise is not just the brilliant idea of anybody in particular. It is something that seems indicated as a result of our new means of travel, our new wealth, our new means of education. In the Peace Corps, you have an educational activity in which there are no subjects and no exams, but in which there is total involvement and total learning. It is the type of education which the teenagers are trying to tell us they want us to produce for them at home, and we don't know how to do it. I'm not sure that we have to, or should, but that is the sort of gripe they are trying to communicate to us.

There's a wonderful book called *The Wheelwright's Shop* which describes the life of craftsman in an English village of two hundred years ago. A wheelwright is a man who, in order to be a simple craftsman making wheels for carts, had to know the needs not only of all the people in the area, but also had to know all the resources, all the kinds of trees, all the kinds of materials available for answering these needs. He developed a sort of rapport with his environment, social and physical, that was totally artistic. *The Wheelwright's Shop* by George Sturt goes on to explain how the wheelwrights would toil incessantly for many hours a day simply because they were so involved in their work, they couldn't be torn away from it. They never needed leisure. They were never bored – like a painter. An artist today is never working. He's doing what he wants to do.

He's playing and he's at leisure at all times, especially when he's working hardest. This is one of the peculiarities of our time. The old world of the job, this is neolithic; the world of the job, little fragmented specialist tasks, is no longer bearable. What people want now are roles that enable them to become totally involved so that they will be totally at leisure, utterly fulfilled. This need for involvement that comes with our new information age is a very mysterious and confusing thing after centuries and centuries of specialism in which there was relatively small involvement.

We have leapfrogged from the age of the wheel to the age of the circuit – the electric loop which feeds back into the user creating the involvement of such loops. One of the effects, for example, is that nearly all the industries of our time are becoming service industries. In the electric age you no longer just produce packages, you produce service. With xerography, for example, the book itself ceases to be a bound package and becomes an electric service, obtainable by telephone and computer from any part of the world, and delivered immediately according to your need and your specifications. Instead of just picking up a book that happens to be on the shelf or on the table for sale, you go and specify your interests, your needs, and that book is then produced by xerography, produced according to your needs, or for your specific interests, in archeology, history, or whatever. So we're moving out of the age of the mass-produced package into the age of the personal and private service.

The great archetype of this sort of situation is education itself. Education is now our biggest industry, many times bigger than General Motors or AT&T. The education industry takes more personnel and more technology and more money than all the other businesses at all. It's a service industry. And many other industries have become service industries in our time. But the function of the teacher is now more and more to save the students time, to accelerate the learning process. The need to accelerate the learning process is connected with the fact that in the age of information there are such vast levels, quantities of information to cope with, that the older methods of filing of classified information in our memories will not serve.

With the satellite and the immediate likelihood of satellite broadcasting to all parts of the world simultaneously, the little round schoolhouse has turned the planet into a form of an artwork. As the satellites go around the planet, the planet becomes not the human habitat so much as an art form, a kind of nostalgic piece of camp, as the popular word puts it.

One of the effects of living with electric information is that we live habitually in a state of information overload. There's always more than you can cope with. Now the young have devised a strategy for dealing with this quite apart from just inattention. They have devised a much more potent strategy of myth-making. Our ancestors in the age of the hunter lived in a mythic world because they had none of the literate means of classification of information. So they formed their information and their traditions into myths for the sake of retrieval, for the sake of restoring insights. A myth is an instantaneous insight that also includes all the stages of a process. The familiar one is about King Cadmus, the propagator of the alphabet, who sowed the dragon's teeth, from which sprang up armed men. This myth indicates not only that he propagated the alphabet in ancient Greece, but also that having done so, and to his dismay, it bred military activity on a huge scale.

One of the reasons that literacy created vast military organizations in the ancient world was that it became possible to control men at a distance by means of written messages and couriers. The alphabet created individualism. It created armies, empires, huge organizations, and the myth of Cadmus tells you all that in a few words. The advantage of the myth is this tremendous compression, this insight and compression.

This sort of change over to an environment of information confuses us on many fronts, not only on the educational front. But the city itself becomes a service industry. "The urb, it orbs." The urb of the city becomes a world form and at the same time becomes a service industry. It becomes an all-at-once service. It becomes a kind of auditory, resonating word. The all-making word is heard again in the world. Harvey Cox has much to say on this in *The Secular City*. The world and the city seem to be re-sacralizing. Having been secular, and fragmented, and specialist for a long time, they suddenly begin to take on this sacred character again like the ancient city. Way back in tribal times the city was a sacred entity, not just a practical, efficient organization.

At the end of the nineteenth century, Frederick Jackson Turner was granted a vision of American history and development, which he called the significance of the frontier in our history. That was in 1890. The electronic telegraph had come into operation a few years before, speeding up events, capsulating the past, and he comes up with the wonderful myth, in the best sense of the word, a great myth of history, understanding the totality of an operation involving many millions of people as single myth, the myth of the frontier.

The telegraph by speeding up information helped to create this new information environment. It led the Danish existential philosopher, Søren Kierkegaard, to become quite alarmed at the prospects of man under telegraph conditions. His *Concept of Dread* was published in the first year of the telegraph, the commercial telegraph, 1844, and he saw the threat of the telegraph, what we now know to have been true, as a threat to human identity. When everything happens at once, when everybody becomes totally involved in everybody, how is one to establish identity? For the past century people have been working at that problem. Quest for identity is a central aspect of the electric age. Naturally, we're looking for identity in the old rear-view mirror where it was before. Perhaps we should be looking for it in corporate institutions instead of in the private sector, as they use the phrase.

Now Turner's mythic approach revolutionized the study of American history by inspiring great controversy and new perception. He provided Harold Innis with the basis for his many studies of the evolutionary stages of our institutions. But even evolution itself becomes meaningless under electric conditions because if everything happens at once, if the DNA particle is programmed from all eternity, or is totally programmed before anything happens, it's an all-at-once operation.

This all-at-onceness of the kind of situation in which electric information involves us is very confusing and disturbing, but at the same time very challenging, very exciting. One of the effects of the same electric world had been to transport the frontier back to Europe. Turner observed that in the settlement of America we have to observe how European life entered the continent, and how American life modified and developed that life, and reacted on Europe. The staples, the furs, the minerals, the wood, the logs that Europe derived from North America had a considerable effect naturally on their social lives. We became John Bull's other gold mine. And this had great effect on the whole of European life. It would take lots of time, and very useful time, to look into.

Rousseau saw the American scene, for example, as the unspoiled natural environment of man, the noble savage environment, the pastoral dream, and this now seems to have reversed. *Bonanza* or the Western, the TV Western, is literally the simplicity where the human spirit can expand in all its original vigor. I'm not promoting these views; I'm simply suggesting that these have been common experiences derived from this kind of world.

The sort of feedback of the American thing into the European situation began to take place in the arts as well as in the economic sector. Henry James spent his entire life portraying the European impact on America and the American impact upon Europe. That was a long time ago. And the kind of situation in which we are now makes it look quite childish and elementary compared to the sophisticated situations we have prepared for ourselves.

1. Francis Bacon, *The Letters and Life of Francis Bacon*, ed. J. Spedding, (London: H. G. Bohn, 1862), vol. 2, p. 99.

2. James Joyce, *Finnegans Wake* (London: Faber & Faber, 1939), p. 598.

Fordham University:
First Lecture
(1967)

In Fall 1967, McLuhan left the University of Toronto for one year to take up an appointment as Albert Schweitzer Professor of the Humanities at Fordham University in New York. During that year, in addition to teaching an undergraduate course in Understanding Media, McLuhan undertook two other projects: to design a new curriculum for elementary schools in light of his theories, and to investigate ways for communications research to work its way into American high schools.

In his introductory lecture on September 18, 1967, McLuhan explains to more than two hundred students registered in his course that war is total education, that television is an X-ray machine, and that the planet becomes a global village because space is reduced by the speed of communication to almost nothing. During his half-hour talk he recapitulates his theory that changes in the communications media are producing a new "invisible" environment with profound implications for society.

I'm very grateful to Steve Allen for having pointed out to me years ago that the funnyman is the man with a grievance. Wherever you find jokes sort of clustering and generating in notable degree, you're going to find a grievance. The ethnic joke is such an area of interface between one cultural group and another which produces endless grievances.

Canada is loaded at present with French-Canadian jokes because French Canada and the rest of Canada have been interfacing and irritating each other over a long period. It tends to get a little hotter from time to time and produces stories like the one about the mouse who was being pursued by the house cat, and he found a little hole in the floor and scampered under, and then after a while all was quiet and then suddenly a kind of "bow wow arf arf" occurred, and the mouse figured the house dog had come along and scared the cat away, so the mouse popped up and the cat grabbed it, and as it chewed the mouse down it said: "You know, it pays to be bilingual." I could tell you some more of those French-Canadian grievance stories. Wherever you find jokes, you find grievances.

This is a nice example of how invisible the real environment always is. We do everything we can to hide from the present because the present is an area in which we feel exposed. We're insecure and, therefore, somewhat unhappy. We do everything possible to conceal the present from us. This is true of men in all periods of human history. They prefer to look in the rear-view mirror. You feel safer back one age. This is just as true of Plato as it was of Shakespeare. Shakespeare was a medievalist in his work. Plato was very much devoted to the old primitive and integral order of tribal man behind him. *Bonanza* is a nice example of where America prefers to live.

Bonanza-land is one stage back, the frontier. It's much safer than the present. But this tendency to always live back one stage puts one in a rather bad position, or the age in a rather bad position, to do any serious navigating.

We happen to live in a time of speeded-up information when we can have access to the past so easily that there really isn't a past. It's all present. The present has become so rich and so complex and terrifying that people do all they can to hide from it. The information environment, for example, of satellites and *Telstar* – what happened with *Telstar* was rather amazing in ways that people haven't noticed. When you put a man-made environment around the planet, the planet becomes an art form. Nature ceases to exist. Nature goes inside a man-made environment. With *Telstar*, with satellites, the natural world, the planet, our external habitat, has become the content of a man-made environment as much as any Hollywood set. In the electric age, there is no more nature. The whole planet becomes programmed like a teaching machine. The whole human environment is now a teaching machine.

James Joyce put it this way: "The urb, it orbs."[1] The city becomes terrestrial in scope and the planet, in turn, becomes a global village. It is reduced by the speed of communication. Space is reduced to almost nothing. This kind of revolution is the one in which all of us are actually living, and it enables all sorts of things to appear and be noticed for the first time that had previously been unobservable. This principle is that, whenever a new technology develops, it creates a new environment for the whole culture, and that environment is totally invisible. This has always been true.

Nobody has ever studied this historically. I'll name a few other wide-open fields of great magnitude that have never been touched by human scholarship. For example, at the present time, one of the effects of the speed-up of information is to bring people so close together that the whole human family becomes unitary in a new way, in a very conscious way. This is tribal, that is, when the whole of mankind becomes conscious of its very close dependence in interdependence, you are moving into tribal conditions of interface and dialogue. The electric age that has been shaping us for the last few decades has been responded to by the teenagers, for example. I call it the TV generation. Those who learned to read and write before TV are a very different bunch from those who learned to read and write after TV. Why? Because reading and writing act as a buffer, a cushion against the X-ray effects of TV. TV is not a picture machine. These snapshots that are going around us belong to the old technology. They're pictures; they're not X-ray. The new technology is X-ray. The kids who have been X-rayed from the cradle go out into the world looking for depth, for involvement. Their twenty-one-year-old brothers and sisters are a completely different bunch from the fourteen-, fifteen-, and sixteen-year-olds. Here's an area for research. Nobody ever researched this, but it's wide-open. When people understand that this generation gap is really a technological gap, it will help them to get things in some sort of order again.

Under speeded-up conditions, mankind is retribalizing in the old technical sense. I'm using the word *tribal* in the new technical sense of living by ear rather than by eye. The word *involvement* has taken over from the word *escapism*. Everybody used to be accused of escapism in the age of the book and the movie, and now they're accused of involvement. The old meaning of cool was detachment. The new meaning is involved. The word *cool* has flipped, and it's a very good indication of what's been going on.

Retribalizing is a technical thing, and I'm not taking a pro or con stand on this or any other of these technical changes. It seems to me rather futile to take a moral stand or make a value judgment on a technological matter. After all, technology is something we did to ourselves. Why become indignant or enthusiastic about a wheel or a telegraph?

Another major shift going on right under our noses and, therefore, invisible, is the shifting of the Western world towards the Orient; that is, we are engaged in Westernizing the Orient by our old technology. This is called Vietnam. Sending the old technology to the Orient to Westernize it – this is the military aspect of things as a teaching machine. The military services are teaching machines, and the whole nature of war as educational is one of the themes I think men have never studied. In war a whole culture goes into action. It's not just specialism. Julius Caesar educated the Goths and the Huns very adequately. He really educated them, and they quickly came back and destroyed Rome. Alexander the Great and Julius Caesar were two of the greatest educators in the history of mankind.

Warfare as a teaching machine, warfare as the whole culture acting as a unified educational service, is never more evident than at the present moment. The educational activity going on in Vietnam is total. It's not specialist. They're not getting courses. They're getting our whole culture by interface. Napoleon educated the Russians in Western ways far more than Peter the Great. He taught them to drive on the right side of the road all through Europe, anyway. Does anyone know if the Russians drive on the right? Well, where Napoleon went, there were technological, military reasons why he wanted the traffic on the right-hand side. He never got to England. They still drive on the left. Never got to Sweden. They're now spending billions putting traffic on the right-hand side. Wherever Julius Caesar went, he taught organization in the Roman, visual, bureaucratic style, laying out straight streets, and so on.

War as education. I think once people realize that war is a major all-out educational effort, they will quickly abandon it as disgusting. People aren't that fond of education. So war has been misclassified. It is actually a teaching machine. The whole culture in action simultaneously equals war. Education simply consists in putting one little bit of the culture in action under controlled conditions – algebra, history – break up the culture into little bits and just allow one little bit at a time and that's education. War is the whole culture in action. Now, as I say, these are utterly unexplored, untouched subjects.

While we are Westernizing the East, the Orient, by our old technology, we are Orientalizing ourselves by new technology. The electric age is giving us all the inner trip. We are spending all our latest billions on Orientalizing the Western world by electric means while spending a much smaller budget on Westernizing the Orient.

Again, anything that is present is invisible, and the fact that we are Orientalizing ourselves, going inward, in depth, giving up all our old superficial ways of external aggression in favor of inner tripping, is quite invisible. I don't know what would happen if people became totally aware of what they were doing in any particular period. I hope we may have something to contribute to this. Awareness is a very desirable thing. I'm not sure if that's a moral judgement; it's just a preference. I can assure you that history reveals that people have never, in any age of the human past, ever known what they were doing, that is, what the effects would be of what they were doing. They've always had verbalized ways of mentioning to themselves what they were doing, but, for example, nobody ever knew what the effect of the wheel would be on the arrangement of human life. Now, in the age of the circuit, the wheel is finished. The motor car is on its very last wheels. You can tell that from all the disturbances and all the miseries created by congestion and density of traffic. The motor car has just about had it in the age of the airplane and the capsule. The wheel in the age of the electric circuit is not a wheel at all.

The motor car has had it. I don't care, that is, I don't hope this is true. I don't wish that it were otherwise, either.

Any technology creates a new environment. It creates a total numbness in our senses because our instincts are to hide from that which is not known, that which is strange, so people are always unaware of the new environment. They are unaware of the electric environment. For example, at the present moment the changes going on in the Negro world are very much tied up to electricity. Nobody knows this. That is, you put a total electric environment of information around a population, and the people who feel it most immediately are those who are closest to the ear world, the total world. The Negro lives a more integral life, a more unified life than the rest of the population. I'm speaking now of populations, not individuals. The general population is more specialized and more fragmented.

The Negro is turned on by electricity. The old literacy never turned him on because it rejected him and degraded him. The old mechanical world rejected and degraded the Negro, but electricity turns him on and accepts him totally as an integral human being. Electricity is organic; it's not mechanical. So the Negro feels that he owns the world under these conditions, as in a certain sense he does. The teenyboppers likewise. They feel exactly the same way about electricity; it turns them on. Literacy doesn't turn them on. It's too specialized. It rejects them. But as soon as the Negro is turned on, he looks to the old technology that had rejected him with anger. The destruction that follows is symbolic. It's not to be measured by some kind of rationalistic method at all. It's the enemy. I mention this as practical. I prefer in a way not to have gotten onto that subject. But I'm merely mentioning it since we're doing this to ourselves. We're creating these electric environments without knowing what they do to people at all, without knowing what the generation gap they create is between child and parent. TV has created a huge gap between children and parents, between children and teachers. And nobody knows why. TV is an X-ray machine. It's not picture. It's pure X-ray. The effect on the inner life of people is fantastic. It's an inner trip. The X-ray generation since TV are completely different people from those before TV. It has nothing to do with the programs they watch. It has to do with the profound experience of being X-rayed, totally, in depth, so that every situation thereafter must have depth. So when a child goes to school after TV and sees the offerings, a little bit of this and a little bit of that, he's absolutely amazed to think that anything so feeble could coexist with anything so magnificent as TV. The outer environment is a great big teaching machine charged with messages, whereas the inner environment of the schoolroom is paltry, feeble, specialist, classified data, like looking up words in a dictionary.

This is the kind of revolution we're living through, and the kind that we have made ourselves, made it happen to ourselves. It is a happening. It's total. I'm simply suggesting that awareness as a strategy has something to be said for it. Awareness of the present, not the future, has a lot to be said for it in preference to our usual habit of looking in the rear-view mirror.

---

1. James Joyce, *Finnegans Wake* (London: Faber & Faber, 1939), p. 598.

Open-Mind Surgery

(1967)

On September 28, 1967, McLuhan gave his first public address as Albert Schweitzer Professor of the Humanities to a business audience at the Hilton Hotel in New York City. The title of his talk, "Open-Mind Surgery," is a play on "open-heart surgery," referring to his theme that the new world of electric circuitry is Orientalizing the population by taking it on an inner trip of "integral, organic involvement."

The computerized universe is a huge development in evolutionary terms, according to McLuhan: "What I'm saying is that if man is now in a position to program the total human environment as a teaching machine for the first time in human history, he is also in a position to exercise some reasonable choice and preference over the programming of that environment. Rational man may really get his first innings out of this computerized universe. Up till now, he's been a kind of little straw blown around by technologies."

I'm very indebted to Steve Allen for an observation he made a long time ago that the funnyman is a man with a grievance. I tried that one backwards and said, "All right, then, where there are grievances, there should be lots of jokes." Mostly it proves that way. In French Canada there are lots of grievances, and you remember how General de Gaulle offered recently to free French Canada. He offered to free the Scots, too. That didn't make the press quite as excited as it should have. He didn't dream of freeing the Irish; I think there would have been a storm. De Gaulle is *à nous, la liberté*. As a result of paying attention to grievances, I've discovered quite a batch of jokes, and I pay special attention to media jokes like the one about the teacher speaking to a class. "Now, class, what does this century owe most to Thomas Edison?" A hand went up and the student said, "Teacher, if it weren't for Thomas Edison, we would be watching TV by candlelight."

The French Canadians revel in jokes like the one about the president of Canadian Shell chatting on the phone with the president of American Shell a couple of years hence and the Canadian president is saying vehemently on the telephone, "We must have a big personnel program in Sherbrooke and a total reorganization of the whole show *tout de suite*." There was a pause, and the American says, "Hey, who the heck you think you talkin' to, white boy."

It was recently that I began to pay a little attention to the effect of radio on the twenties. As you know, it brought in the jazz age and the jazz babies. It brought in Hitler, that is, Hitler is a man who played it by ear, who was a retribalized man who had a new message to offer to the Germans by way of their regaining their tribal identity. The other night the CBS show on Germany was terrifying. It revealed that the Germans are desperately seeking for a new tribal identity. It was Ray Bradbury who pointed out, and maybe others too, but I just happened to hear him say that violence is the quest for identity, and whether you do it *à la* John Wayne, or *à la* Negro riots, whether you do it individually or corporately, the quest for identity can only be satisfied by violence. It doesn't have to be a punch in the face. It can be violence one does to one's own nature. Self-discipline, asceticism, and so on are also forms of violence. But the twenties, having become alienated from the visual world of goals and simple directions, began to live it up. They got turned on, and they began to live it up in the jazz age. That turned the Negroes on. It was one of their great moments, not as great as their TV moment by any means.

The coming of this new tribal mentality to the 1920s inspired Gertrude Stein to point out to her age group that they were all a lost generation. This is the condition of the twenty-ones and over right now, a lost generation because they were too old to tribalize, to get the jazz baby message of involvement in tribal form. The twenty-ones and over today are much too old to get the TV tribal message of involvement. They are, and we are, a lost generation, in that sense expendable.

To come back for a moment to the 1920s, when those jazz babies reached the job plateau ten or fifteen years later, they arrived with this total involved habit, without any specialized directions or preferences of the old-fashioned John Wayne variety, a man who knew exactly what he wanted. When you put the tribal man in a decision-making area, he quickly reveals his loss of direction, and you've got a slump, literally. When the man of visual direction and specialism is pushed aside by the electronic, all-around, turned-on, swinging person, "I got plenty of nothin'." Remember *Porgy and Bess*? That was the great message of the twenties. When those people reached the job plateau, we had a slump. But it's nothing compared to the slump that we're just about at in another five years. The slump that's coming from the TV generation will be on a much greater scale unless we decide to operate.

The German world recovered direction and identity by violence, by war, and got into action and recovered itself. I suppose the same war helped us to recover our sense of direction in spite of the jazz babies. But this kind of use of the airwaves, of the total human environment as a teaching machine, a programmed teaching machine, is what it has become. This extension of our own nervous system as a total environment of information is, in a sense, an extension of the evolutionary process. Instead of its taking place biologically over many thousands of years, it has happened to us in the last decades. It is now possible to traverse many millions of years in seconds by putting evolutionary extensions of ourselves outside as environments, as teaching machines. The man-made environments that are now planetary are, in terms of evolutionary development, a greater step than anything that ever happened to our biological lives in the whole biological past.

Just because they are environments, they're invisible. This is a peculiarity that's the meaning of the message to the fish. Or we don't know what discovered water, but we're sure it wasn't a fish. One thing that is always invisible to occupants is the environment. What stands out loud and clear is the old environment, the rear-view mirror, so to the TV operator, the movie world is very visible. TV is invisible. The TV world is an X-ray world; TV is an X-ray device. The viewer of TV is being X-rayed at all times. The radiation goes right through him. The children get this message at once, and they carry the message out into the schoolroom, where they find no resemblance to TV X-ray, where subjects are laid out in little classified forms, algebra, history, and such, no depth, no involvement. So our children, our teenyboppers, and the under-twenty-ones are utterly confused by the discrepancy between the environment they encounter as entertainment and the environment of the schoolroom.

This confusion on a bigger scale is repeated in the Vietnam business. Vietnam is getting the Western treatment. It's getting a thoroughly massive educational program. We put our whole mechanical, industrial environment around it as an educational machine. This is what Julius Caesar did with the Huns. Alexander the Great did. War as an educational institution is the most powerful because it deals with the whole environment in action as an educational teaching machine. Somebody or other gets the message. What we're doing in Vietnam is using the old nineteenth-century machine program for the Orient. What we're doing to ourselves is Orientalizing ourselves by our own latest technology. Instead of the outer trip, the inner trip. Instead of individual specialism, integral, organic involvement. The TV world and the world of electric circuitry automatically ensure the Orientalizing of the whole of the human population exposed to it.

When the Negro experiences our new electric environment, he is turned on as never before. Black power just surges up as he feels this involvement and congenial hospitality of the electronic environment, and he looks around and he sees the old literate, mechanical environment, which had always rejected him, degraded him, alienated him, and this naturally enrages him. He destroys, if possible, the enemy.

India, China, and Africa will do this on a much bigger scale as soon as they encounter an electronic environment. Up till now they are getting the nineteenth century, not the twentieth. The nineteenth century was the old machine age of fragmentation, literacy, and specialism, the one that the liberal mind has never got beyond. It's never studied the later environment. It's literate, visual, detached, non-involved. The visual man is the only detached human being that ever lived on this planet. All the other senses except sight – smell, hearing, touch, movement – are involving; they are discontinuous and unconnected. Only the electronic permits a total encounter with the discontinuous and the disconnected.

The world of discontinuity came in most vividly with the telegraph and the newspaper. The stories in the newspaper are completely discontinuous because they are simultaneous. They're all under one dateline, but there's no story line to connect them. TV is like that. It's an X-ray, mosaic screen with the light charging through the screen at the viewer. Joyce called it, "The charge of the light barricade."[1] In fact, *Finnegans Wake* is the greatest guide to the media ever devised on this planet, and is a tremendous study of the action of all media upon the human psyche and sensorium. It's difficult to read, but it's worth it.

Now we are moving our evolutionary process out into the environment itself instead of putting something inside; Darwin thought of man as inside an environment; it never occurred to Darwin that it would be possible to program the environment itself as evolutionary. Darwin is a literate, nineteenth-century man. He had no intimation of the electronic information circuit or total human environment of information. Teilhard de Chardin is a little more aware of the meaning of the electronic than Darwin. But this Darwinian obsession with man's thumbs, fingers, and appendages as biological evolution is a perfect example of the rear-view mirror. Right in the midst of the great evolutionary step into the electronic circuitry, Darwin remains obsessed with this content of this old environment. What we are saying is that if man is now in a position to program the total human environment as a teaching machine for the first time in human history, he is also in a position to exercise some reasonable choice and preference over the programming of that environment. Rational man may really get his first innings out of this computerized universe. Up till now he's been a kind of little straw blown around by the technologies.

One of the reasons why we can now notice these environments generated by new technologies is that they yield to one another so rapidly that it's almost impossible not to notice the changing of the scene or the guard. But no previous human age ever understood the effect of these human endeavors on the environment as they affected that age. They always understood the effect on the previous age, but never the effect on themselves. The history of Utopias is the history of rear-view mirrors. Every Utopia is a picture of the preceding age, like *Bonanza*-land. There is never a Utopia that copes with the present, any present. Man seems to have this built-in device of rear-view-mirrorism because he fears this confrontation with the total human environment. The only person who seems to habitually encounter and report on the new present is the artist. Whenever he passes in his new report, he is at once branded as a kook because the present is always unrecognizable because it is not safe to look at.

When one understands the operation of these new environments on our human equipment, it then becomes easy to predict and forecast the changes that will take place on many levels of programming, even of entertainment. The end of the star system is automatic with TV. The end of baseball as a star system, like prizefighting, won't work for TV. It's cool. It's involving. It doesn't permit the flashy, blazing, hot star to operate. The kids today are not interested in stars. The Smothers Brothers are not stars in the old sense at all. The kids are very interested in the techniques used by the Smothers Brothers. They're very interested in the processes, how things are done, how things happened and do happen. This following of complex processes is part of the involvement need of an electric age.

Expo in Montreal has been a great success for the simple reason that they pulled the story line off the whole show, like a Fellini movie, no story line. A story line hots it up and leaves the spectator outside, looking on helplessly without involvement. Expo is a mosaic without connections, just like page 1. Total involvement results. Everybody wants to go back again and again to follow through this or that pavilion's activities. And Expo is exactly like the TV screen; it's a mosaic, it's not a picture. There are no pictures on TV. Even when you put a movie on TV, those are not pictures; it still comes through as a "charge of the light barricade." It's an involving, mosaic situation in which the story line or connected spaces are not there. The viewer has to supply them; that's how he gets involved. However, I'm not trying to present any programs of action or directions of action. I'm merely trying to suggest ways of perceiving the situation we have developed for ourselves.

I've recalled another communication story about the two Navajo Indians who were having a little chat across an Arizona valley by smoke signal. Midway through their chat, the AEC [Atomic Energy Commission] released an atomic charge, and when the big mushroom cloud cleared away, one of the Indians sent up a little smoke signal to the effect, "Gee, I wish I'd said that."

The world of communications stories is very rich. A sort of old-fashioned one is about the two goats that were chomping on an old batch of films that had been thrown out behind the MGM studio in Hollywood. One of the goats kicked open an old tin that held a picture of *Gone with the Wind* and signaled to his companion to come and try some of it. When his pal came along and nibbled a little of *Gone with the Wind*, the first one said: "How did you like that?" After a little meditation, the other one said: "Oh, I think I like the book better."

Media stories can be very useful, though, for studying media. I heard one about the two mice in a nose cone, moving around the planet. One says to the other, "How do you like this kind of work?" The other says, "Oh well, better than cancer research." You can see the grievances sticking out a mile. Good old Steve Allen and his observation. If jokes arise from grievances and irritation, then perhaps they serve as a release or a catharsis of the same.

I have this uneasy feeling that I've released a number of these atomic charges, predicting that the next depression is not far off once our TV kids reach the job plateau. They don't want jobs. The next phase is the dropout executive. The reason for that is very simple. A big exec doesn't have a job; he has sixty jobs; that's a role. A mother doesn't have a job; she has forty or fifty jobs; that's a role. In the electronic age, where all activities become associated and inter-related, you can't have jobs; you can only have roles. So the whole job structure is a story line or an organization chart that has to be yanked out if big business is to survive. The organization chart is as dead as baseball, and for the same reasons, one thing at a time. Baseball can't live in a TV age because it's one play at a time, one pitch, one hit, one catch. It won't work. In an all-at-once world, one thing at a time won't work on the job front or on the sports front or on the educational front.

I'm trying to be helpful here; I'm not trying just to put on an act. I wish somebody had told me some of these things a few years ago. It's been exciting to discover them, but I also feel there's a bit of waste. For example, apropos the TV generation, there ought to be studies made – and heaven knows we're trying to do a few of them this year – of the difference between those who saw TV after they learned to read and write and those who first began to watch TV before they learned to read and write. I think you'll find there's a big watershed there. Two totally different kinds of human beings are coming on, oncoming traffic, two totally different kinds. The kids who learned to read and write before they saw TV had a kind of immunity, a cushion for their senses that made the TV impact very much less. I think those exact dates and age groups ought to be carefully ascertained and studies made because the whole future programming of work and education depends on it.

Vietnam is our first TV war. That's why people won't buy it. It's too involving. All the previous wars were fought on hot media like movies, pictures, photographs, and press, but now people experience war as something that involves them profoundly, and they don't want to have anything to do with it. It isn't a question of trying to ascertain the pros and cons, the rights and the wrongs of the Viet Cong or whomever. That has nothing to do with it. It's just this new phase of total involvement and participation that is untakable. The new gang coming on, the kids who are really turned on, will find that war or any other war much less tolerable than the present generation. No comparison here.

We do waste an awful lot of time assigning fanciful reasons for these events and selecting very unfortunate and wasteful goals for our energies when we do, after all, have a rather deep desire to work out a livable equilibrium for the human community. We do go to so much trouble to create violent upsets and disequilibrium. But I'm not sure how successful a reprogramming could be in our lifetime.

I'm just going to check to see whether I've omitted any crucial funny stories. Probably not. I see one hint here. Columbus went too far. Or a similar observation, Plymouth Rock should have landed on the pilgrims. In our present world there cannot be a Columbus. External exploration has ended, and the future of exploration is necessarily internal, whether in medicine or in entertainment. So since I feel you are in good humor at this moment, I think I will fold my tent like the Arabs and quietly steal off.

---

1. James Joyce, *Finnegans Wake* (London: Faber & Faber, 1939), p. 349.

TV News as a
New Mythic Form
(1970)

On a hot summer day in August 1970, Tom Wolfe and Marshall McLuhan were filmed by TVOntario in conversation in the garden of McLuhan's home in Wychwood Park in central Toronto. Wolfe had been an acquaintance of McLuhan's since the mid-sixties, when he wrote a seminal article about him in the Sunday Herald Tribune Magazine called "Marshall McLuhan. Suppose he is what he sounds like, the most important thinker since Newton, Darwin, Freud, Einstein, and Pavlov – What If He Is Right?" (November 21, 1965).

As a conversationalist, Wolfe describes McLuhan as amazingly adept at integrating whatever subject anyone could suggest to him into his theory on the spot without a moment's hesitation. Woodstock, Sirhan Sirhan's assassination of Robert Kennedy, the Middle East war of 1967, the war in Vietnam – it did not matter. McLuhan was never at a loss. When Wolfe asks him how a pathetic person like Sirhan Sirhan could become a heroic figure, McLuhan attributes it to the myth-making power of the news media.

McLuhan: It's grand being able to sit down and talk about some of the things that we're interested in. And I'm intrigued by the fact that you're from the South. I've always had an interest in the South. My wife is from the South.

Wolfe: That's right, she's from Fort Worth.

McLuhan: I'm interested in the fact that you were a Southerner, and this relation to an oral tradition is a great advantage to a literary man. In the twentieth century it's very remarkable that all the best writing has come out of Ireland or the American South because of this close relation that the English language has to the spoken word in those areas. And this, too, seems to have something to do with the existence of jazz and rock as art forms that, without an oral tradition of corporate public address, this kind of music would not occur.

Wolfe: I'm sure a lot of it also has to do with preaching.

McLuhan: There again is a public address system.

Wolfe: I really can't think of any part of the country where preaching among both black and white preachers has had such prominence, and where people get so fulsome in their expression, whether it's the very stilted kind of speech that the Southern Episcopalian minister uses with the *uh* expressions, forgive us *uh* our *uh* our trespasses and *uh*. It's kind of an English mannerism.

McLuhan: Those hesitations and those intervals are actually very involving. It makes the audience just hang on the next phrase. It's like a stutterer who keeps you on the ropes waiting for him to form another word.

Wolfe: This whole idea of an oral tradition certainly did come back with the Beat Generation. In fact, I think the main contribution that people like Allen Ginsberg and Gregory Corso and Ferlinghetti made was to break up the academic poetry which had become so strong after the Second World War, and which was very formal, rigid.

McLuhan: Which kind of poets did you have in mind by the word *academic*?

Wolfe: Practically everyone who was reviewed in the *Kenyon Review*.

McLuhan: I see. That would include all the Southern poets and the Irish poets and some of the British poets. But the establishment, the poetic establishment you meant that had built up around the academic study of Pound and Eliot and Yeats?

Wolfe: I think the people who took off from Pound and Eliot and who were writing after the war really neglected the oral side of someone like Eliot or Pound, and were more in love with the fact that someone like Pound was filling his work with scholarly allusions and mythical allusions. And it became a totally –

McLuhan: Mythic study, and very anthropologically oriented, archeologically oriented, very learned. But the fact is those poets themselves came out of the age of radio and would have been unthinkable without that radio ground around them. And the whole of the English language took on a tremendous new oral life from radio. I'm sure that this was carried straight on into jazz and rock music.

Wolfe: A lot of poets, in England, particularly, were being read on the air, weren't they?

McLuhan: And recorded, and there was a great disk presentation of poetry in the thirties especially.

Wolfe: This has become even truer today when so many poets make their living – it's the only way they can make a living – going around to universities and giving readings. And I really think it's done a great deal for poetry.

Tom Wolfe and McLuhan.

McLuhan: On the other hand, there has been a demand for this. The public wants to confront the poets. Surely the poets have been writing for radio. There have been all sorts of poetry plays and poems and songs written for radio and for TV. Why don't we just consider the tremendous publics opened up to what formerly had been rather limited publics with the written or printed forms, the tremendous new publics opened up by television and radio for poetry and drama and stories.

Wolfe: The big mistake people writing for those media make is that they don't collect what they've done for television or radio and put it in a book because once it's in a book, it exists for the first time in the official sense. I discovered for myself I must have written 110 magazine articles, but once I collected them into a book –

McLuhan: They take on a totally different character. Well, they're more portable. The magazine is expendable and disposable. The book is still retained. The pocket book, of course, tends to suffer the fate of the magazine. It gets handed around from person to person and tossed into corners and so on. But you feel that writers often get lost just in the media. They toil away as scriptwriters for film, for TV, and for radio, and their work doesn't show.

Wolfe: It disappears into the ether, and then they become very frustrated.

McLuhan: But by going to campuses and reading to big publics, they recapture their relationship to their audience.

Wolfe: I think the poets do. I also think it improves their writing because they know they're going to have to read it.

McLuhan: They have to make it sound good out loud. Do you know that there is very little prose that you can read out loud? Most prose has never been written for the ear at all. When you have to suddenly read it in public, a quote, you suddenly realize you're reading the work of a man who never heard what he was writing.

Wolfe: A lot of dialogue is very convincing in novels, but when someone tries to put it on the screen, it's terrible. I imagine that Dickens might suffer that fate.

McLuhan: I'm not sure. There have been quite a number of movies made of his work. I suppose they've been adapted rather than just repeated. But there is a real oral tradition behind Dickens. He's not a literary man and only became respectable as a literary figure in the thirties and forties. In fact, I don't think that until television did the English critics accept Dickens as a serious literary force. He was farce, not force, up until TV. I'm sure the future of the writer is not exactly the nineteenth-century future. But I'm sure that it's bigger than anything the nineteenth century ever dreamed of.

Wolfe: Well, writers still have a few exclusive tricks up their sleeves. So far film, whether it's television or in the movies or theater, has never managed to use point of view. They've tried everything. They've tried voiceover. They've tried acting as if the actor, as if his eyeholes were a camera so he could only see himself in a mirror and that sort of thing.

McLuhan: Point of view requires stasis and that's not involvement. So these new media demand that you get involved and become part of the action. The TV image is not like the film image, a simple snapshot. It is an actual live vortex in action, and is behind what we call the happening. TV itself is a kind of happening, technically, and it tends to involve people in its own vortex. And what is called in the new journalism the happening, what Norman Mailer calls the novel as history and the history as novel, is a kind of use of the total environment as a surround or as a vortex of action in which everybody, the reader and the characters, are all involved. His visits to the Chicago Democratic Convention are nice opportunities in which to show vortex.

Wolfe: I think that's where most of the important changes, certainly during the 1960s, took place. I think a hundred years from now historians, that's assuming that the Chinese will have any interest in our character or history, won't look at the 1960s in the case of, say, the United States as the era of the war in Vietnam, of the moon shot, or anything of that sort. I think it will be looked at in terms of what you refer to as the ground has changed, the way people have changed their ways of living.

McLuhan: We used to concentrate on figures, and now the ground itself has become figure. The area of attention has shifted from the older characters to the ground. Now that includes audience. The audience has now become actor. Don't you think this is a tendency as a result of developments in our time?

Wolfe: Well, certainly Woodstock was a perfect example of it. Woodstock is probably the great, typical event of our times because –

McLuhan: Instant city.

Wolfe: It was set up. From the very beginning there was going to be a movie made of Woodstock. As it started out, every one of us was paying our eighteen dollars for the weekend. Gradually, so many people came, they just abandoned that and let them all come in. But actually they should have paid them all eighteen dollars as they came in because they became the show. And Ken Kesey and his group, the Merry Pranksters, their whole idea when they started these acid tests in California where they would get five hundred people in a hall and give everyone LSD and have these lights, was there would no longer be any separation between the performer and the audience. Kesey himself said that he was no longer going to write novels, that he was tired of being a seismograph. He wanted to be a lightning rod.

McLuhan: Consider in that regard what we call coverage. Coverage now is no longer just on a single individual but on a whole complex action. In turn, don't you think that both in Vietnam and in the North of Ireland, the audience wants to get into the action, that the coverage encourages the audience to get into the action? I have been told by reporters from the North of Ireland that when the news is not on, and the cameras are ready to go, the public is all out in the streets, ready to go into action as soon as the cameras are.

Wolfe: That's marvelous.

McLuhan: They all retire inside to watch the news, and then come outside to participate in covering the news and in acting it out themselves. Now the difference between hired actors and the public itself is tending to merge. Isn't that what you're saying? This kind of unexpected flip. It happened in the Eichmann trial. The coverage pushes up the figure into heroic dimensions dramatically, but at the same time involves the audience so completely in the process of his action that they begin to feel far more guilty than he did. He appears merely as a person carrying out orders, the orders of the community. He was a well-adjusted, nice guy who was doing what had to be done, according to the audience command, the audience being so involved in this process that it now begins to feel like a villain. Therefore, they want to cut that show right out of their lives. But the happening, I think, is of this kind. It is a situation in which the audience and actor become one, in which audience becomes actor, not spectator. So in Truman Capote's *In Cold Blood* the audience becomes actor.

Wolfe: One reason that *In Cold Blood* made such a splash was that you had a novelist, whose reputation had just been drifting downhill for about seven or eight years, suddenly turning to non-fiction in a novelistic manner, completely rescuing his standing, and, in fact, becoming a much bigger figure than he had been before, becoming a kind of international celebrity as a result.

McLuhan: Would you regard *In Cold Blood* as a kind of documentary, a reconstruction of actual events?

Wolfe: It is very much like a documentary. Most crime stories done in non-fiction are reconstructions, and they have the strengths and weaknesses of reconstructions. One of the great weaknesses is that the dialogue is seldom very good because no one can ever remember it that well.

McLuhan: Or invent it that well.

Wolfe: I reconstructed a great deal of the material in *The Electric Kool-Aid Acid Test*, but in that case it was dealing with a group of people, Ken Kesey and his Merry Pranksters, who had been absolute technological freaks. They were obsessed with the idea of recording their lives in every possible way. They kept tape recorders running all the time. They even used videotapes. They used tape-lag mechanisms. They took movies of their own lives. They kept diaries. They had strange diaries in which you couldn't write in your own diary; only other people could write in your diary.

McLuhan: This was a kind of recycling of themselves.

Wolfe: And I think very rapidly, too. One of their ideas was to get very high on LSD, and to have videotapes running which you could play back in about a minute or two, and the interval between what they had actually done and what they would now see on a screen was so brief, and they were so high, that it was as if all the lags in life were being overcome.

McLuhan: It's like the playback in modern sports. This is surely one of the greatest art forms of our time, the instant playback, the instant replay which concentrates attention on the actual process of the game. I've asked various footballers what's this done to your actual play, and they tell me we've had to change the play, open it up a bit so that people can see it as if it were replay.

Wolfe: They use this in teaching tennis now. There are all these academies – I'm sure they make a lot of money, too – they get all sorts of people together in groups of eight and ten, and they have the videotape running, and after a person takes a few whacks at the ball, they bring them into a room and show them what they were doing. But think of applying this to everyday life.

McLuhan: I think anybody who has heard his voice on a recording is appalled at the condition of his vocalizing, and the same when he sees himself on TV. I think he resolves that he never wants to look at that or like that again. In other words, I think the need the media creates is for acting, that people realize that just their plain private self is not adequate to the media, and I think that this drives people towards dramatizing.

Wolfe: And yet it's very hard to dramatize yourself on television, wouldn't you think?

McLuhan: I don't know. We're trying to do it right now.

Wolfe: I'm not feeling very dramatic. I know that you've been thinking a lot recently about the idea of the put-on, with several meanings. And I believe I heard you say there were no great North American symphony conductors because of . . . and I never got that straight.

McLuhan: It's not a direct or simple thing but it began with my discovery that Americans go outside to be alone and inside to be with people. In all other cultures, Hindu, Russian, Japanese, English, European, people go outside to be with people. They socialize outside, they sit outside, they eat outside and talk and visit in cafés, in pubs and bars, and so on, and they go home to be strictly alone. They don't invite strangers into their homes. In America, on the other hand, there is no privacy in the American home, and strangers are often entertained and welcomed. But when an American goes outside, he is his private self. He does not put on a mask. Other people, when they go outside, tend to put on masks in speech or in dress or in actions. I suggest that there are no American symphony conductors, Leonard Bernstein having grown up in his early years at least in Vienna, that an American conductor, no matter how much he knows about orchestras or publics, is not able to dramatize himself in the way of putting on the whole orchestra as his private mask or corporate mask, nor is he able to put on the whole public as a mask, whereas a European doesn't even think about this as a problem. When he stands up in front of the orchestra, he ceases to be a private person and becomes an actor instantly. Now an American not only does not like acting or putting on a public, but he does not put on a corporate or standard voice when he speaks. He uses his private voice, and this, of course, enables the Americans to have no class consciousness and no class structure.

Wolfe: This would explain why they behave differently in restaurants or hotels.

McLuhan: And when they're in a movie or at a restaurant, they want to be alone. They don't want advertisements in movies because they go there with their dates to be alone, whereas Europeans will put up with conversation in movies and advertisements in movies and so on. They do not go out to be alone. This then led to this whole problem of the put-on. The comic, when he gets up in front of people, puts people on by twisting their arms. The jokes that comics tell are the ones that cut very close to the bone, close to home, and unless the jokes come close to home, they don't have any relevance or put-on value. So the ethnic jokes are the ones that are closest to home and the areas where things hurt, where there are abrasive contours and interfaces and so on. So most jokes tend to have this minority quality or irritation quality, just as games people play often tend to have a rather destructive and violent quality. But a put-on, therefore, tends to be a way of hurting the public. It's an act of aggression against people. And I think that a writer, when he picks up his pen, has to put on his public that way. If he has something to say, it's going to hurt.

Wolfe: I remember your predicting once that if the coverage of the war in Vietnam was suddenly withdrawn, the war would grind almost to a halt. And this happened during the Six Day or Seven Day war in Israel and Egypt, and suddenly all those correspondents for wire services that were concentrated in Saigon were suddenly summoned to go to the Middle East. And it was true that during that week the fighting, which was hot at that time, suddenly just came down. Unfortunately, it was not a long enough interval.

McLuhan: I think it might be worth an experiment – people are always talking about the need to understand media by experimentation – to turn it all off for a week, that is, there would be no newspapers, no radio, no TV, no telephones, nothing for one week. What do you think would happen? How would people respond to a complete blackout?

Wolfe: Well, in Ken Kesey's phrase this would be a media fast. He wrote an open letter to *Rolling Stone* magazine in which he said: "I am breaking a six-month media fast in order to . . ." I like that idea of the media fast.

McLuhan: Well, when there is no news as, for example, in a prison camp where all news is cut off, there have been studies of what happens, an enormous outburst of rumors. The oral thing takes over, and people generate incredible rumors about what's happening, what's going to happen, what's being planned, and so on. So probably there is a kind of rumor control by means of coverage even when it's completely fake. That is, most news is literally fake because it has to be made, then selected, and the very, very tiny bits that are actually written up and reported and presented to the public are fictions in every sense of the word, aren't they? They are fictions in the sense that they do not correspond to actually what is going on, but they are made, literally, created.

Wolfe: Do you think this explains the really strange fascination that Arthur Bremmer had with Sirhan? He obviously looked at Sirhan Sirhan as some kind of heroic figure. He wasn't this poor, helpless, useless human being who had done this desperate thing, certainly not in Arthur Bremmer's eyes.

McLuhan: No, and again he had made the news. Sirhan had made the news. Now, this you can take in every sense of the word as having gotten into the news, having been created into a vast figure by the news. "Making the news" is a very strange phrase, but the media themselves can now create events that are so much bigger than people, so much bigger than the audience, that it really is a new mythic form. The coverage of the Vietnam war is done by more people than those who are actually fighting in Vietnam. The numbers of people covering Vietnam business around the world, and participating in it through newscasts, the numbers are many many millions, and so the war then becomes a fiction, a colossal fiction. This then leads people like Clifford Irving, just as much as Bremmer, into thinking that someone like Howard Hughes can be turned into a myth, a genuine fake.

Wolfe: Politicians very quickly learn that if you want to get on the six o'clock news at night, if you want to get on NBC, you do something in Cleveland before noon, because if you don't do it before noon, the network does not have time to process it and to fly it to New York to be shown nationwide. It has to be something of great magnitude to happen at three o'clock and make the six o'clock news. So you have this marvelous spectacle of politicians all over the country having a press conference at 10:30 a.m. in Cleveland, and another one is having one at 10:30 a.m. in New Orleans, because, being so aware of all this, they very quickly catch on to it. And commercial enterprises have caught on to the fact that all the networks want the last item on the six o'clock news to be funny. I don't know exactly why this is true. Maybe it's because you've just been slaughtered for half an hour, and they want a little kicker so the last item on the news every day is always really a covert advertisement, pure and simple. The last one I remember seeing in New York before leaving was a group of models in bikinis dressed up in ape suits who went to the Central Park Zoo and in their ape suits they fed the monkeys bananas. And every single local news station in New York ran this thing as the last item on the show, and it was an advertisement for the *Conquest of the Planet of the Apes.* And so it's not so much the networks manipulating the news; it's just setting up the system and then other people use it.

McLuhan: This raises something we really haven't talked about, that is, the news has to be a put-on. It has to put on a public, and so the last comic twist in the news would be a way of dismissing the public and saying "so long." You know how in *Laugh-In* they do it that way. They have a rapid roundup, at the last as a dismissal, of wildly improbable comic turns. Well, Tom, I think there are a whole lot of wonderful topics we've managed to broach at least, but it would take a long time to get into them.

Wolfe: I would really like to run down a checklist of all the predictions you made six years ago that people thought were absolutely crazy that have come true. I remember your saying once that there would be a time when they would have to pay students to go to school. Well, there is a new federal program in the United States where they're doing just that. They figure the only way they can get these large numbers of kids interested is to give them a little salary.

McLuhan: I've always been very careful never to predict anything that had not already happened. The future is not what it used to be. It is here. And when you look into the rearview mirror what you ordinarily see is not the car you passed but the truck that's coming up on you fast. Never look back. They may be gaining on you. So you can't lose. You can't win. The present includes the past and the future.

The Future
of the Book
(1972)

In London, England, on April 27, 1972, a conference took place at the National Film Theatre to debate the subject "Do Books Matter?" Chaired by the Duke of Edinburgh and jointly promoted by the Working Party on Library and Book Trade Relations and the National Book League, the meeting was designed to bring together some of the most brilliant minds in the Western world to question the whole value of books in a modern technological society.

McLuhan was invited to contribute as an international figure and world expert on the significance of mass media in an electronic age. He spoke on the topic of "The Future of the Book," a theme of particular interest to him as a literary scholar. Contrary to the common misconception that he thought the book was dead, his speech reveals that he envisions the book on the verge of radical new developments: "Today there arises the possibility of direct brain-printing of books and data, so that the individual can be equipped instantly with all he need ever know. Such a bypassing of all reading raises many questions about the function of books."

Much confusion about the figure of the book, past, present, and future, results from the new ground which surrounds both book and reader today. The printed book is a definitive package that can encode ancient times and be sent to remote destinations. More than electronic information, it submits to the whims of the user. It can be read and reread in large or small portions, but it always recalls the user to patterns of precision and attention. Unlike radio and phonograph, the book does not provide an environment of information that merges with social scenes and dialogue. As the levels of sound and video images envelop the user, he turns off in order to retain his identity. The first video age presents the example of a generation of literate people who in various ways have turned off or gone numb as an involuntary tactic to preserve private individuality. In contrast, the merely tribal man, or preliterate, would seem to feel no threat to his personal life from the new electronic surround.

If the book can be sent anywhere as a gift or as a service environment, the paradoxical electric reversal is that it is the sender who is sent. This flip or chiasmus of form and function occurs at the level of instant speed and is characteristic of telephone and radio and video alike. When the *ground* or surround of a service assumes this instant character, the *figure* or the user is transformed. Thus the book does not have its meaning alone. The book in the preliterate world appears as a magical form of miraculously repeated symbols. To the literate world, the book serves a myriad of roles ornamental and recreative and utilitarian. What is to be the new nature and form of the book against the new electronic surround? What will be the effect of the microdot library on books past, present, and future? When millions of volumes can be compressed in a matchbox space, it is not merely the book but the library that becomes portable.

There are many ways in which the book and literate values are threatened by the mere fact of the electric service environment. The extreme distraction presented by the acoustic and cinematic rivals of the book brings decreasing opportunities for attentive and uninterrupted reading. Beginning with the typewriter, and then the mimeograph, the nature of the book underwent immense change of pattern and use. The typewriter changed the forms of English expression by opening up once more the oral world to the writer of books. Whether it was Henry James dictating interminable sentences to Theodora Bosanquet, or the semi-literate executive giving abrupt letters to a secretary who could spell and compose grammatically, the new services of typing and mimeographing transformed the uses and the character of the printed word.

In tracing some of the effects of the printed word on liturgy, James F. White points to a result of the mimeograph:

A further development considerably affected Protestant worship though not mentioned in any liturgical textbooks. In 1884 a Chicago businessman, A. B. Dick, solved a business need for rapid duplication by inventing a process for stencil duplication. It proved so efficient that he marketed it under the name of "Mimeograph." Gutenberg made it possible to put prayer books in the hands of people; Dick made prayer books obsolete. Prayer books are mostly propers which are hard to locate and confusing to most people. Dick gave each minister his own printing press and a new possibility of printing only what was needed on any specific occasion. Xerox and other processes promise to do the same for hymnals. These developments have simply completed what Gutenberg began, and in worship as elsewhere we are now flooded with printed paper.[1]

It would seem clear from this passage that the advantage of always studying any *figure* in relation to its *ground* is that unexpected and unheeded features of both are revealed. It is in Ray Bradbury's fantasy fiction *Fahrenheit 451* that the world ahead is shown to fear the book as the cause of dissension and diversity of opinion and attitude. As such, the book is the enemy of unanimity and happiness; therefore, it must be destroyed. To save the book from the furious fireman and the incinerator, numerous individuals volunteer to memorize separate works as a means of perpetuating them to a life beyond the flames. Today there arises the possibility of direct brain-printing of books and data, so that the individual can be equipped instantly with all he need ever know. Such a bypassing of all reading raises many questions about the function of books.

The future of the book raises the question of whether men can ever program their corporate social lives in accordance with any civilized pattern by any other means than that of the printed book. There is no question that people can associate in large numbers without books or training in literary perception. At present even computers depend on their programming for literate people. The entire use of yes-no bit programming is from the alphabetic modes of Western civilization. But many people look to the computer to bypass present forms of human action and limitation. Yet, even the written and printed word, it might be argued, has helped us to surpass ordinary human scale. Eric Havelock's *Preface to Plato* and *Empire and Communication* by Harold Innis have shown us how Western man has shaped himself by the phonetic alphabet and the printing press. Paradoxically, the very individuality achieved by these means has inhibited the study of the effects of technology in the Western world. Having fostered a high degree of private self-awareness by literacy, Western man, unlike Oriental man, has shunned the study of technological impact on his psyche. No Western philosopher has evolved an epistemology of experience or looked into the relation between social and psychic change in regard to man's own artifacts. So, it is only fitting that Western man should be excluded from awareness of the effects of his own action by the principal effect, namely, his own private psyche. That prime product of our own phonetic literacy is the shaping awareness of individual interiority and privacy. Before our time, any approach to study the breaching of this interior life by external technological action has been repulsed by Western man. Since the fifth century B.C., neither Plato nor Aristotle nor any subsequent investigator has regarded the creation of Western individuality by the action of phonetic literacy. Likewise, the effects of any outer technology whatever on the inner life of man have been avoided until the electric age.

But today the electronic effects are so massive that they cannot be ignored. Students such as Eric Havelock and Harold Innis have looked into the matter and found the personal transformations by technology to be quite demonstrable. Writing or printing or broadcasting constitute new service environments that transform entire populations. And whereas private, Western man shuns and deplores the invasion of his privacy by the historian or by the psychologist, the tribal or corporate man feels no such reluctance in checking the psychic and social effects of technology. He feels no such compunctious workings of nature as does the civilized man. The Oriental societies have always been eager to know the effects of any technical innovation on their psychic lives, if only in order to suppress such innovation. Western private man prefers to say, "Let's try it and see what happens."

The future of the book is inclusive. The book is not moving towards an omega point, but is actually in the process of rehearsing and re-enacting all the roles it has ever played, for new graphics and new printing processes invite the simultaneous use of a great diversity of effects. *Poésie concrète* has inspired many new uses of older printing methods and has called for the invention of new print and paper surfaces. Photo-printing permits the imposition of letters on and through many materials. Print can be moved through liquids and impressed upon fabrics, or it can be broadcast by TV and printed out on plastic sheets in the home or study or office. Thus the current range of book production varies from the cultivation of the art of the illumination of manuscripts and the revival of hand presses, to the full restoration of ancient manuscripts by papyrologists and photographic reproduction. Taking xerography alone, we find the book world confronted with an image of itself that is completely revolutionary. The age of electric technology is the obverse of industrial and mechanical procedure in being primarily concerned with process rather than product, with effects rather than content.

In the present age of ecology it is not easy to write off the future of the book apart from the effects of the book, for ecology is concerned with anticipating effects with causes. In order to program any situation it is necessary to know exactly what components are congruous and which ones are incongruous with the intended effect. The printed book has a very different meaning and effect even for different age groups in our Western world. It certainly has a very different effect in the Orient from its effect in the West. Today, for example, the meaning of the book for young people in their TV environment is exotic indeed. The printed book, by its stress on intense visual culture, is the means of detachment and civilized objectivity in a world of profound sensuous involvement. The printed book is thus the only available means of developing habits of private initiative and private goals and objectives in the electric age. These characteristics do not develop in the cultural milieu of electric sound and information, for the acoustic world, like the "auditory imagination" defined by T. S. Eliot, is not private or civilized but tribal and collective.

The book has always been the vortex of many arts and technologies, including speech and mime and pictorial elaboration. At first, the printed book seemed to have excluded much of the richness of the manuscript. At first, many buyers of printed books took them to the scriptorium to have them copied out by hand. For one thing, the printed or mass-produced book discouraged reading aloud, and reading aloud had been the practice of many centuries. Swift, silent scanning is a very different experience from manuscript perusal, with its acoustic invitation to savor words and phrases in many-leveled resonance. Silent reading has had many consequences for readers and writers alike, and it is a phase of print technology which may be disappearing.

Gutenberg had, in effect, made every man a reader. Today, Xerox and other forms of reprography tend to make every man a publisher. This massive reversal has, for one of its consequences, elitism. The nature of the mass production of uniform volumes certainly did not foster elites but rather habits of universal reading. Paradoxically, when there are many readers, the author can wield great private power, whereas small reading elites may exert large corporate power.

Today, Xerox has brought about many reversals in the relation of publics and writers, and these changes help us to see not only the past and present but also the future of the book. For example, Xerox extends the function of the typewriter almost to the point where the secret, personal memo is moved into the public domain, as with the Pentagon Papers. When notes for briefing individuals or groups are first typed and then Xeroxed, it is as if a private manuscript were put in the hands of the general reader. The typewriter plus photocopying thus, unexpectedly, restore many of the features of confidential handwritten records. Contemporary dialogue in committees depends very much on this new Xerox service; but the very public character of the service is difficult to restrict. The result is that confidential briefing is now beginning to take an oral rather than a written form. It seems useful to consider the impact of Xerox if only because it illustrates how profoundly one technology can alter traditional patterns of relation between writing and speaking.

To write about the present of the book, with a keen eye on the changing *ground* for the book as *figure*, is to realize how many new forms the book has assumed even in our time. In all patterns, when the *ground* changes, the *figure*, too, is altered by the new interface. When the cinema and gramophone and radio and TV became new environmental services, the traditional book began to be read by a completely different kind of public. If Gutenberg created a new kind of human being with new perception and new outlook and new goals, the electric age of radio and video has perhaps restored a public with many of the oral habits of the pre-Gutenberg time.

Had anyone asked about the future of the book in the fifth century B.C., when Plato was beginning his war against the poetic establishment and its rigorously trained rhapsodes, there would have been as much confusion and uncertainty as now. The time when the world of Nature would appear as an extension of the glorified art of the scribe lay ahead. For Plotinus, the stars are "like letters forever being written in the sky, or like letters written once and for all and forever moving."[2] Concerning the seer, Plotinus says that his art is "to read the written characters of Nature, which reveal order and law."[3] Yet, strangely, in ancient Greece "there is hardly any idea of the 'sacredness of the book,' as there is no privileged priestly caste of scribes."[4]

Reading and writing were assigned to slaves in ancient Greece, and it was the Romans who promoted the book to a place of dignity. But essentially, "It was through Christianity that the book received its highest consecration. . . . Not only at its first appearance but also through its entire early period, Christianity kept producing new sacred writings."[5] It was the elucidation of these writings that called forth encyclopedic programs of learning and scholarship which drew as freely on the page of Nature as on the *sacra pagina* of Revelation.

Throughout the Middle Ages, the metaphor of the Book of Nature dominated science. The business of the scientist was to establish the text and its interpretation by intensive contemplation, even as Adam had done in the Garden where his work had been the naming of creatures. Unexpectedly, the massive and ancient trope of the Book of Creatures ended with printing. It became old cliché. What would a sage have said in the early Gutenberg time if asked to predict the future of the book? How would Erasmus or Cervantes have answered this question just at a time when the printed book was opening new vistas of reputation and influence to writers? A century later, Francis Quarles [1592–1644] could still play with the idea of the Book of Nature in a merely decorative way:

> The world's a book in folio, printed all
> With God's great works in letters capital:
> Each creature is a page; and each effect
> A fair character, void of all defect.[6]

Shakespeare had still found vitality in the trope of the Book of Nature in *As You Like It* where the banished Duke in Act 2 finds the voices of nature:

> And this our life except from public haunt,
> Finds tongues in trees, books in the running brooks,
> Sermons in stones, and good in everything (Act 2, Scene 1).

It was Mallarmé, the symbolist, who proclaimed that "all the world exists to end in a book."[7] His perception complements and also reverses the ancient and medieval tropes of the Book of Nature by assuming that in both the industrial and electric ages Nature is superseded by Art. Thus the future of the book is nothing less than to be the means of surpassing Nature itself. The material world is to be etherealized and encapsulated in a book whose characters will possess all the formulas for the knowledge and recreation of Being. Such was the ambition of James Joyce, whose *Finnegans Wake* is a symbolist summa involving all creatures whatever. Joyce embraces both art and artifact in his encyclopedia of creatures, encompassing his task by means of language alone.

Taking the book in the more mundane sense of a printed package, it can have as many incarnations as it can find new techniques to wed. I have already alluded to the power of Xerox to transform the reader into publisher. Indeed, the prominent American publisher, William Jovanovich, has written about the subject of how reader and writer and publisher switch roles today:

> William Saroyan wrote me from Paris:
> "I seem to have the notion that anything anybody writes has got to be published so that the writer can begin to feel better, I suppose . . ."
> I replied: "Your idea that anybody who writes should be able to be published may, in fact, come true. Xerography is a process that can make this possible, but whether it will make people feel better I cannot surmise, unless they happen also to be Xerox shareholders. Certainly if publishing becomes universal, and if it is regarded as a kind of civil right, or a kind of public requital, then our concept of literary property must change. Everything will be published and it will belong to everybody – power to the

people. There is nothing illogical in your idea. If everyone finds a publisher, he will then find a reader, maybe just one reader – the publisher himself. Of course, writers want lots of readers, but this desire will be less and less fulfilled as there are more and more writers. Quantity declines as specialization declines. Eventually, every man will become at once a writer, publisher, librarian, and critic – the literary professions will disappear as a single man undertakes all the literary roles.[8]

In the light of this new publishing technology, it is less surprising to hear the alarm in the voice of Jean-Paul Sartre, who had already anticipated the Jovanovich report in *What Is Literature?*

> From this point of view, the situation of the writer has never been so paradoxical. It seems to be made up of the most contradictory characteristics. On the asset side, brilliant appearances, vast possibilities; on the whole, an enviable way of life. On the debit side, only this: that literature is dying. Not that talent or good will is lacking, but it has no longer anything to do in contemporary society. At the very moment that we are discovering the importance of *praxis*, at the moment that we are beginning to have some notion of what a *total* literature might be, our public collapses and disappears. We no longer know – literally – for whom to write.[9]

With Xerox, publishing enters on new courses in a way that is reminiscent of the fantasies entertained by early printed writers such as Cervantes, Rabelais, and Montaigne.

I would like to conclude with some observations that the printed book inspired in Montaigne in the sixteenth century. From his wonder and aspirations in that early printing time, we may well derive some assurance to begin anew the exploration of the present, when the tape recorder and the videocassette afford the writer new roles and new publics for the published word. When the book and its author can mount the back of another medium like radio or TV or satellite, the scale of the operation both in time and in space seems to abolish the difference between the microscopic and the macroscopic. And to quote Montaigne:

> Amusing notion: many things that I would not want to tell anyone, I tell the public; and for my most secret knowledge and thoughts I send my most faithful friends to a bookseller's shop . . .[10]

If Erasmus or More or Tyndale had been asked about the future of the book, could they possibly have imagined that it was to become the means of intense inward investigation of the private psyche on the one hand, and the creation of huge reading publics on the other hand? The interplay of these two factors created in the mind of Montaigne a new feeling of moral obligation:

> I owe a complete portrait of myself to the public. The wisdom of my lesson is wholly in truth, in freedom, in reality . . . of which propriety and ceremony are daughters, but bastard daughters.[11]

He is saying here that the "self" demanded by the revelation of the printed form throws into the discard all traditional decorum and propriety. What is called for is the minute realism of self-study:

> He studies himself more than any other subject; that is his metaphysics and his physics. "I would rather be an

authority on myself than on Cicero. In the experience I have of myself I find enough to make me wise, if I were a good scholar." Fortune has placed him too low to keep a record of his life by his actions; he keeps it by his thoughts. He makes no claim to learning: "I speak ignorance pompously and opulently, and speak knowledge meagerly and piteously."[12]

By the same token, if print drove Montaigne to minute self-investigation and self-portrayal, may we not expect the book of the electric age to turn this perspective on patterns of corporate human energy and association? The videocassette offers an immediate opportunity for the reader and the author to enter into a totally new relationship. The reader will be given an opportunity to share the creative process in a new way, indicating that the book is on the verge of totally new developments.

1. James F. White, *New Forms of Worship* (New York: Abingdon Press, 1971), p. 28.

2. Ernst Robert Curtius (tr. W. R. Trask), *European Literature and the Latin Middle Ages* (New York: Pantheon Books, 1953), p. 307.

3. Ibid., p. 308.

4. Ibid.

5. Ibid., p. 310.

6. Ibid., p. 323.

7. Stéphane Mallarmé, *Œuvres complètes* (Paris: Librairie Gallimard, 1945), p. 378.

8. William Jovanovich, "Universal Xerox Life Compiler Machine," *American Scholar* 40 (1971), p. 249.

9. Jean-Paul Sartre (tr. Bernard Frechtman), *What Is Literature?* (New York: Philosophical Library, 1949), p. 241.

10. Donald Frame, *Montaigne: A Biography* (New York: Harcourt, Brace, & World, 1965), p. 82.

11. Ibid., p. 291.

12. Ibid., p. 253.

The End of the
Work Ethic
(1972)

On November 16, 1972, McLuhan addressed a luncheon meeting of four hundred members of the Empire Club of Canada at the Royal York Hotel in downtown Toronto. The Empire Club is an organization made up of Canada's conservative elite in economics, academia, and business, and McLuhan specifically targets this group with his topic, "The End of the Work Ethic."

For big business to survive under electric conditions of the new information environment, it has to totally change its job structure. "The Protestant work ethic," McLuhan notes, "as characterized by private goal-seeking and self-advancement is meaningless in a world of electric information." Pattern recognition is the new form of work which combines into one the roles of hunter, engineer, programmer, researcher, and aesthete.

I have a friend on the faculty of the University of Toronto, Professor John Abrams, who went to Russia a few months ago. Before he started out, I asked him to pick up a couple of authentic Russian stories for me. I'd like to know just what sort of stories they tell in Russia. In due time, he relayed a couple to me. One story concerned a yokel visitor to Moscow who was a bit of a provincial, who was oohing and ahing over this wonderful city and thanking God for our wonderful subways and our wonderful economy and our wonderful stores and our wonderful people. His acquaintance remonstrated: "You don't thank God, you thank Stalin." And the visitor said: "Sure, but what if Stalin dies? Then you really can thank God."

There's an undercurrent of grievance in all good jokes, as you may have noticed. Jokes are mostly concerned with grievances, and when new kinds of jokes appear, you can depend upon it, there is some sort of uncomfortable abrasive area of tension developing in the community. The Newfie jokes represent a certain amount of mild irritation relating to group maladjustment. Most of us have heard some of the good news/bad news jokes that project a sort of gripe about the rise of the new subjective journalism and the disappearance of the old objective journalism. I'll have a chance, perhaps, to refer to examples later. The old journalism had attempted to be objective and detached. For many years this meant giving both sides at once – the pro and the con, the yes and the no, the for and against, the black and the white. When you give both sides at once, you are in the tradition of objective reporting. If something is about to be done in the community, you explain what's against it and what's for it. On the other hand, the new journalism is an attempt to give all sides at once by simply immersing you in the total situation, Norman Mailer–style, or Truman Capote–style, as *In Cold Blood*. Norman Mailer's *Miami and the Siege of Chicago* is not so much a report about the convention or the candidates or the policies of the parties. Rather, Mailer plunges us into a happening for the feel of it. The new journalism is subjective immersion. Whether it's fact or fiction doesn't matter any more. When you get all sides at once, it is a sort of fiction anyway. So, Norman Mailer's subtitle on his *Armies of the Night* is "History as a Novel – The Novel as History," pointing to the loss of boundaries between reporting and fiction. When we say "He made the news," we point to the fact of news as fiction, as something made. Perhaps that is why there is always just enough news every day to fill every paper with nothing leftover.

Now, I am going to turn to my topic at once so that I will not be accused of not having gotten around to it. "The End of the Work Ethic" is quite directly related to the fact that under electric conditions of the new information environment, it is no longer possible to have a mere goal. When everything happens at once, when you are inundated with information from every quarter of the globe at once and every day, you cannot fix a distant objective and say, "I'm going to move toward that point." That point is already in rapid motion as you are, and long before you take a step in that direction, everything will have changed.

The work ethic, insofar as it meant private goal-orientation, is not practical and disappeared some time ago. Related to this situation is change in the job. The job will no longer hold up; it's too specialized an action for the electric age. The job as specialism, as a fixed position in an organization chart, will not hold up against the simultaneous jostling and the interfacing of simultaneous information. What is taking the place of the job is role-playing. When you are moonlighting and starlighting, that is role-playing; and most people are doing this in some degree or other. The job-holder drops out as the consultant drops in.

Role-playing, in effect, means having more than one job. A housewife is a role-player because she has many jobs, and so do farmers and many other people in the community. They are not essentially job-holders but role-players. Notice the stress on playing. I am going to bear down on that theme a bit. Electrically, we are moving into an age of play which will bring many new patterns of work and learning. There is about to appear from a Canadian publisher a book called *The Leisure Riots* [by Eric Koch]. It concerns a rebellion of a community against a group of executives who are trying to transform fun and games and leisure-time activities into industrial packages. When skiing or hockey or football or tourism become encumbered with equipment and consumer packages and are made expensive and difficult, they cease to be fun. The play has gone. This is a situation which is reflected in various aspects of our world at present. It is another area of grievance. The idea of play and recreation has gained much new meaning in this age, not only from such classic studies as *Homo Ludens* by J. Huizinga, but from quantum mechanics. Huizinga relates the play principle to the development of all our institutions, so that the very idea of "party government" implies the interplay of diverse attitudes and policies as a means of equilibrium and social adaptation. In chemistry and new compounds, the very fact of the chemical bond has become the resonant interval of interfacing particles. Matter is now seen to be constituted by resonating intensities of interplay minus any connections whatever.

All are familiar with the play between the wheel and the axle as the very principle of mobility, and we seek to avoid the uptight on one hand, or the too slack, on the other hand. But it could be argued that the dropout is a victim of the uptight situation and that he drops out in order to regain touch. When the wheel and the axle get too close, they too lose touch. When they are too distant, they collapse. To be in a bind is to lose touch as much as when we become too remote. When a group of sane people volunteered to enter mental institutions as if they were in need of therapy, the staff was unable to detect their sanity, but the seriously deranged pointed at the intruders at once, saying, "They're playing." Without play, there looms the shadow of madness in private as in corporate life, whence the danger of specialism and bureaucracy when the rigors of classification overlay the relaxed countenance of harmonious faculties. The artist is always at leisure because he must keep his mind at play, and he is never more at leisure than when seeking the solution of tough technical problems.

Let me refer to an item from *Business Week* of last month – "The Case Against Executive Mobility." The theme is quite simple, and relates to the recognition that frequent transfers to new homes and places of work have become a sterile way of life in North America. To pull up stakes and move families as a means of advancing up the corporate ladder is no longer considered a very practical or rewarding thing by big business. Another observation in the same way belittles a chance for significant job promotion that would affect the emotional lives of school-age children. At the other extremes, local IBM operations have become so large that executives can often get the effects of diversified managerial experience they need for advancement without moving anywhere. To move people around quickly is to cease to have a community.

The current bestseller, *Jonathan Livingston Seagull*, is a kind of sideswipe or spoof on the Protestant work ethic. If the words *Jonathan Livingston* suggest the great medical missionary of the last century, the word "Seagull" is rich in aerial and spiritual suggestion. In fact, J. L. S. is a very hard-working and very aspiring individual whose endeavors are richly rewarded by rapid advancement aided by delightful skyscapes

and photographs. We read how J. L. S. never ceases from prac-
ticing all manner of flight techniques: "A hundred feet in the
sky he lowered his webbed feet, lifted his beak, and strained
to hold a painful hard twisting curve through his wings."[1]

Of course, the idea of elevation and levitation are easily
associated with the strenuous and precise maneuvers of the
seagull. However, he soon phases himself out of all polarity in
the company by the sheer speed of his success.

Since a seagull never speaks back to the Council Flock,
Jonathan's voice was raised in exhortation: "Irresponsibility?
My brothers! Who is more responsible than a gull who finds
and follows a meaning, a higher purpose for life? For a thousand years we have scrabbled after fish heads, but now we have
a reason to live – to learn, to discover, to be free! Give me one
chance, let me show you what I've found . . . The flock might
as well have been stone."[2] But that didn't stop him from
moving up the organization tree very fast into another dimension altogether where he encountered another voice: "How do
you expect us to fly as you fly? You are special and gifted and
divine, above other birds. Look at Fletcher! Lowell! Charles-
Rowland! Are they also special and gifted and divine? No more
than you are, no more than I am. The only difference, the very
only one, is that they have begun to understand what they really
are and have begun to practice it."[3]

Now the toil for self-advancement and self-perfection is
one that belongs to the work ethic, allegorically spoofed in this
book. Perhaps the success of the book has been that the reader
can take it both ways at once, either as a critique or a pan-
egyric of the old work ethic.

To return to the theme of the obsolescing of the work ethic
in a world of electric information, it is helpful to understand
that the public today has taken on a new role. Because of the
very simultaneity of electric information and programming,
there really are no more spectators. Everybody has become a

member of the cast. The consumer function itself is outmoded. On Spaceship *Earth* there are no passengers, but all are crew.

Just how quickly the public participation flips from audience to actor has recently been noticed in Watergate, and earlier in the trial of Lieutenant Calley. When audience participation becomes too extensive, the villains on trial flip into public heroes because the audience quickly identifies with them, especially when they are men who are carrying out robotlike activities which necessarily resemble those of the average audience. When the audience identifies with a villain on trial, the public at first begins to feel villainous and guilty itself, and then resolves the problem by flipping the villain into the role of a cultural hero. Another situation in which the opportunities for public participation are very dramatic has been discovered by the hijackers of airplanes and other public services. One of the principal causes or motives for hijacking is precisely the desire for coverage and public attention on the part of these unhappy performers. So attractive is this form of notoriety and coverage, that the reporting of these events has had to cease.

Perhaps we can get into the situation of today via the remark of the little boy on his first airplane ride. Once they had taken off, he said: "Daddy, when do we start to get smaller?" The little boy's question is rather complex since it is plain that the plane gets smaller, while the cabin does not. The little boy would never have asked such a question in an open-cockpit plane, for not only does the plane get smaller, but the occupants also feel increasingly insignificant. However, the enclosed space of the cabin of the plane presents a very special kind of structure, namely, a visual space, that is, a space or *figure* without a *ground*. Visual space has peculiar properties that are not shared by spaces created by any of the other senses. Visual space is continuous, uniform, connected, and static, whereas the sense of touch, of hearing, or of smell is discontinuous, disconnected, non-uniform, and dynamic. Such is also the case with acoustic space, the space in which we all live in the electric age. To a considerable degree, Western literate man in the nineteenth century lived in visual space which he thought of as normal, natural, rational space. With the advent of a world environment of simultaneous and instantaneous information, Western man shifted from visual to acoustic space, for acoustic space is a sphere whose center is everywhere and whose boundaries are nowhere. Such is the space created by electric information which arrives simultaneously from all quarters of the globe. It is a space which phases us out of the world of logical continuity and connected stability into the space-time world of the new physics, in which the mechanical bond is the resonant interval of touch where there are no connections, but only interfaces.

Many people may still suppose that these matters belong to the realm of speculation and abstruse scientific investigation, but the present fact is that we all live in this new resonating, simultaneous world in which the relation between *figure* and *ground*, public and performer, goal-seeking and role-playing, centralism and decentralism, have simply flipped and reversed again and again. Civilized man, Euclidean man, whose visual faculties were sharpened and specialized by Greco-Roman literacy, this kind of aggressive, goal-seeking, one-way entrepreneur, has simply been dislodged and put out of countenance by the new man-made environment of simultaneous electric information. It is important for survival to understand that the simultaneous data of the omnipresent information environment is itself structurally acoustic. When people understand this acoustic structure as their new habitat, they will at once recognize the risks for the strange goings-on in the human psyche and in human society in the effort to relate to this new habitat. It goes without saying that any of our senses adjusts at once to any change in levels of light or heat or sound, and so it

is with the changes in the environmental structure that is constituted by the information services of our twentieth-century
world. Instantaneous retrieval systems and data processing
entirely alter the nature of decision-making. The old job-holder,
secure in his niche in the organization chart, finds himself an
Ishmael wandering about in a chaos of unrelated data.

Let me try to make this matter more vividly comprehensible
by relating a story which concerned my own first discovery of
acoustic space. A group of us which included Carl Williams
(now president of the University of Western Ontario), Tom
Easterbrook (Political Economy Department, University of
Toronto), and Jacqueline Tyrwhitt (Architecture and Town
Planning at Harvard) were discussing the newest book of
Siegfried Giedion, *The Beginnings of Architecture*. Jacqueline
Tyrwhitt (who had worked with Giedion on this study for
years) was explaining how Giedion presented the fact that the
Romans were the first people to enclose space. The Egyptian
pyramids enclosed no space since their interior was dark, as
were their temples. The Greeks never enclosed any space, since
they merely turned the Egyptian temples inside out, and a
stone slab sitting on two columns is not an enclosed space. But
the Romans, by putting the arch inside a rectangle, were the
first to enclose space. An arch itself is not an enclosed space
since it is merely formed by tensile pressure and thrust.
However, when this arch is put inside a rectangle, as in the sec-
tions of a Roman viaduct or in the Arc de Triomphe, you have
a genuine enclosed space, namely, a visual space. Visual space
is a static enclosure, arranged by vertical planes diagonally
related. Thus, a cave is not an enclosed space any more than is
a wigwam or a dome (Buckminster Fuller's domes manifest the
acoustic principle rather than the visual or enclosed space principle). The wigwam, like the triangle, is not an enclosed space
and is merely the most economical means of anchoring a vertical plane or object.

At this point, Carl Williams, the psychologist, objected that the spaces inside a pyramid, even though dark, could be considered as acoustic spaces, and he then mentioned the characteristic modes of acoustic space as a sphere whose center is everywhere and whose margins are nowhere (which is, incidentally, the Neoplatonic definition of God). I have never ceased to meditate on the relevance of this acoustic space to an understanding of the simultaneous electric world. The basic structural fact about simultaneity is that the effects come before the causes in such structures, or, the *ground* comes before the *figure*. When the *figure* arrives, we say, "The time is ripe."

Living electronically, where the effects come before the causes, is a rather graphic and vivid way of explaining why distant goals and objectives are somewhat meaningless to "neuronic" man. Electronic man, that is, works in a world whose electric services are an expansion into environmental form of his own nervous system. To such a man it is meaningless to say that he should seek or pursue distant goals and objectives, since all satisfactions and objectives are already present to him. This explains the mystery of why preliterate and acoustic peoples appear to us to be so deeply satisfied with such shallow resources and means of existence. Acoustic man, living in a simultaneous environment of electric information, is suddenly disillusioned about the ideal of moreness, whether it be more goods or more people or more security or more fame. Acoustic, or electronic man, understands instantly that the nature and limits of human satisfactions forbid any increase of happiness through an increase of power or wealth. Acoustic man naturally "plays it by ear" and lives harmoniously and musically and melodiously. Ecology is only another name for this acoustic simultaneity and the sudden responsibility for creating ecological environments pressed very suddenly upon Western man on October 4, 1957. That was the day when *Sputnik* went into orbit, putting this planet inside a man-made environment for the first time. As soon as the planet went inside a man-made environment, the occupants of the planet began to hum and sing the ecological theme song without any further prompting.

When the planet was suddenly enveloped by a man-made artifact, Nature flipped into art form. The moment of *Sputnik* was the moment of creating Spaceship *Earth* and/or the global theater. Shakespeare at the Globe had seen all the world as a stage, but with *Sputnik*, the world literally became a global theater with no more audiences, only actors.

Another theme with which we are acquainted today is that of the dropout, whether it be Agnew Agonistes or the school-kid who cannot see how he can relate to the curriculum fare. The dropout is literally trying to get in touch. This paradoxical fact is obscured to visual man, always looking for connections and unable to see any rationale in the dropout process. Earlier I had mentioned the example of the interval of play between the wheel and the axle as exemplifying the very nature of touch. To keep in touch is always to maintain a resonant interval of play between *figure* and *ground*, between our jobs and our lives, and between all of our interests and responsibilities. The paradoxical fact that touch is not connection but an interval, not a fixed position but a dynamic interface, helps to explain the changes from the work ethic of yesterday to the diversified role-playing demanded of us in the new electronic environment. Archie Bunker is extremely popular as exhibiting much of the discomfort that results from his being frequently subjected to new situations which are alien to his specialized emotional rigidities and his very particularized points of view that result from his having a fixed position at all times when looking at his world. Time was when a man's point of view was thought of as something integral and important. Today, to have a fixed position from which to examine the world merely guarantees that one will not relate to a rapidly changing world. The problems in the Watergate hearings are notable exemplars of this new situation. Their expertise and highly specialized skills as advisers to the government bureaucracy seemed only to qualify them for involuntary show biz. They were a group of back-room confidants who were suddenly flushed out into the public eye as expendable.

Wyndham Lewis, the painter-writer, once said to me, "The artist is engaged in painting or describing the future in detail because he is the only person who knows how to live in the present." However, to live in the present, or to take today now, involves acceptance of the entire past as now and the entire future as present. Science fiction doesn't really measure up to the everyday reality because the real has become fantastic. The future is not what it used to be, and it is now possible to predict the past in many scientific senses. With carbon-14 tests available, we can now predict why we shall have to rewrite most of the past, simply because we can see much more of it simultaneously. At electric speeds of information it is not only the assembly line that is outmoded by computer programming, but the organization chart that has been outmoded. On the one hand, we hear about new difficulties at the Ford and similar plants in getting workers to come full-time. Two or three days a week seems to be as much as the Detroit worker feels necessary to be involved in gainful employment. Absenteeism means that most of the cars now turned out on Fridays and Mondays are duds. This results from having to staff the assembly lines at those periods of the week with substitutes who have perhaps never seen an assembly line before. The story is told in Arthur Hailey's *Wheels*.

The instantaneous-simultaneous programming which has succeeded the one-thing-at-a-time world of the assembly line is familiar to the IBM world. It is now possible to include all the assembly-line programs of an entire enterprise on a few little solid-state chips. Along with the flip from hardware sequence

speeds comes the flip from hardware scale to software programming. The shift from hardware quantity to software is
nowhere more spectacular than in the microdot library con-
trived by Vannevar Bush for the use of the astronauts. He made
it possible to include, in the space the size of a pinhead, 20
million printed volumes in retrievable form.

One of the peculiarities of the reversals in this new world
of the electric information environment is the return of the
mentality and the figure of the hunter on a massive scale. Man-
hunting, whether under the mode of commercial or military
espionage, is one of the biggest businesses of the twentieth
century. It is a world in which the vision of Edgar Allan Poe
and the work of Sherlock Holmes merge to become a new kind
of work. It is a kind of aesthetic work of pattern recognition
far removed from the Protestant work ethic of goal-seeking.
Today the hunter, the engineer, the programmer, the researcher,
and the aesthete are one.

Intimations of this coming change occurred as early as
Carlyle's *Past and Present* (1843), in which he contrasted the
world of his time with the life of a medieval monastery in
which work and prayer combine to create community. As specialism and industrialism developed in the nineteenth century,
the artists combined to confront and denounce the anti-
humanism of this new mechanical world of fragmentation.
Paradoxically, the aesthetes, from Ruskin and Pater to Oscar
Wilde, agreed that art and work must blend once more to
create good art and the good life. The medievalism of the
Pre-Raphaelites was not nostalgically motivated so much as
concerned with the need of their time to recover an integral
work and life pattern. Comically enough, medievalism has
come upon us in the sixties and seventies in the so-called hippie
costume of international motley which we associate with the
dropouts. Shaggy hair and shattered jeans are worn by those

who are "agin" the establishment, even as motley was the clown costume of the rebel against authority. International motley is not limited to any continent, nor did it originate in any theory or concept of dress. It is as spontaneous a thing as country music or Bob Dylan's enunciation. The figures of emperor and clown, of establishment and anti-establishment, represent an age-old conflict. Strangely, the clown is a kind of PR man for the emperor, one who keeps the emperor in touch with "where it's at," regaling him with jokes and gags which are frequently of a hostile intent.

The nineteenth-century revolution had been, in part, to substitute the laws of the market and the economy for the laws of nature. Economists and sociologists sought to discover the Newtonian laws of the universe embedded in the marketplace. The *Sputnik* event was another thing altogether which simply obsolesced the planet itself as nature disappeared into an art form. The moment of *Sputnik* was the moment of creating Spaceship *Earth* and the global theatre which transformed the spectators into actors. Today, therefore, everybody demands a positive participation in the world process. This, of course, is one of the marks of Women's Lib. Whereas the suffragettes had merely sought to gain the right to vote, women today sense that they are totally involved in the social process on a non-specialist basis and want a large piece of the action. Watergate has shown how the top executives of any big operation are extremely vulnerable today because the most confidential operations can be submitted to public scrutiny on a mass scale. The Watergate cast represented a very specialized group, so much so that not one of them was able to provide any examples of decision-making. However, they were able to dramatize the plight of the specialist at high levels who, in effect, has nothing to do with decisions. At electric speeds nobody makes decisions but everybody becomes participant in a complex situation for which he can take no responsibility whatever.

Another peculiar feature of this electric time is that people not only do not make decisions individually, but in terms of the movement of information, it is the sender who is sent. It is the user of the telephone who goes to Peking and back, and so it is with TV or radio. When you are "on the air," you are everywhere at once. This is a power beyond that of the angels, according to Thomas Aquinas, for they can only be in one place at a time. This is one way of pointing to the revolution in the work ethic, since we haven't a clue as to how to adjust our traditional lives to this kind of instant transportation of whole populations. Moreover, it means that the information environment automatically involves everybody in the work of learning, for the user of a radio program, or a newspaper, or an advertisement, is assisting the community process as much as anybody in a classroom or on an assembly line. When you are watching a play or a ball game, you are working for the community. We live in a world of paradoxes because at electric speed all facets of situations are presented to us simultaneously. It used to be the specialty of "the Irish bull" to do this. For example, a recent example mentions an exchange between two chiropodists. One says: "I have taken the corns off half the crowned heads of Europe." That particular one contains several contradictory facets and metaphors, but as a wag said: "A man's reach must exceed his grasp, or what's a metaphor?"

In the sixteenth century, Gutenberg made every man a reader, creating vast new publics, and in our time Xerox has made every man a publisher, creating (via position papers) vast numbers of big committees. When "Everyman" becomes a publisher, the office boy can give the Pentagon Papers to the world. Such papers are really position papers that may never have been read by anybody. On the other hand, from the point of view of the publishers, Xerox is a total invasion of copyright because all books go into the public domain via this new kind of access.

One of the dominant effects of our electric time is the effect of speedup on decision-making and on awareness of innumerable patterns and processes which had been quite undetectable at slower speeds. In fact, electric speed is tantamount to X-ray in relation to all human activities, invading privacy in both the personal and in the political sectors alike, and creating new patterns of involvement and participation in all affairs.

What is called the new journalism is, in effect, immersion reporting in contrast to the old objective reporting. The old and new journalism, corresponding to the old nineteenth-century hardware and the new electric software and the old and new politics, match these accordingly. The old objective journalism had aimed at giving both sides of the case, whereas the new immersion journalism simply plunges the reader into the experience of being on the scene, or being part of the scene, in "you are there" style. "You are there" is merely another name for movie and TV experience, but the old ideal of objectivity in reporting, of giving both sides at once, now appears in retrospect as an illusion. That is, to give the pro and the con is worlds away from being objective, since it is necessarily a point of view at a distance. In the days of objective reporting, it was always taken for granted that the news behind the news or the inside story was necessarily quite different from the outside or objective view of the situation. The outside story was fit to print and the inside story was not fit to print, in the style of conscious versus subliminal investigation of human psychology. It was Freud who began the immersion approach to the human psyche and the reporting of the subliminal or inside story of human motivations. Personally speaking, my own approach to media study has always been to report the subliminal effects of our own technologies upon our psyches, to report not the program, but the impact of the medium upon the human user. Surprisingly, this kind of reporting of the hidden effects of media creates much indignation. Many people would rather die than defend themselves against these effects. The corresponding flip from objective to immersion techniques in politics presents itself in the form of a political shift from parties and policies to images and services. That is, political parties and their explicit policies have simply been obsolesced by the images presented by the party leaders on the one hand, and the services taken for granted by the community, regardless of the party that happens to be in power, on the other hand.

The drastic flip from objective to immersion reporting has spawned a new genre of jokes which contrast good news and bad news in a style represented by such stories as that of the doctor who reports to the patient: "I have some bad news for you. We cut off the wrong leg. But there is some good news. The withered limb is beginning to show some signs of life!" Or the story of the master of the group of galley slaves, who says: "Men, I have some good news for you. You are going to have an extra noggin of rum today. But now the bad news – the captain wants to go water skiing!" And in even briefer form – Othello says to Desdemona: "I have bad news for you. I'm going to strangle you. Now I'll tell you the good news. I found your handkerchief!"

The revolution I have been describing in reporting and in politics is the theme of a book called *Deschooling Society* by Ivan Illich. The theme of the book is simply that since there is now more information outside the schools than inside, we should close the schools and let the young obtain their education in the general environment once more. What Illich fails to see is that when the answers are outside, the time has come to put the questions inside the school rather than the answers. In other words, it is now possible to make the schools not a place for packaged information, but a place for dialogue and discovery. This new pattern is recorded in the observation that twentieth-century man is a person who runs down the street shouting, "I've got the answers. What are the questions?" There are various versions of this observation, some of them attributed to people like Gertrude Stein. At any rate, when information becomes totally environmental and instantaneous, it is impossible to have monopolies of knowledge or specialism, a fact which is extremely upsetting to nearly everybody in our present establishments. The loss of monopolies of knowledge and specialism is recorded in many fun books like *Parkinson's Law* and *The Peter Principle* and *One-Upmanship*. In fact, it's a basic principle that when new grievances occur, new jokes come with them.

Another basic change resulting from electric speed is the shift from centralized to decentralized structures in every sector of community life. At the political level it is called separatism, but it has been happening on a huge scale in business and in education as well. In private life it is called dropoutism. In fact, that is the theme of my book *Take Today: The Executive as Dropout*. The work ethic is being overlaid very quickly by these new forms of organization so that twentieth-century man not only experiences his subliminal life being pushed up into consciousness, but the daily process of living takes on an increasingly mythic or corporate participation in processes that had previously been kept down in the unconscious. It was Harold Innis in his essay on "Minerva's Owl" (prompted by his studies of the Canadian economy) who showed how the ordinary technologies of everyday life have effects upon us that are in no way dependent upon the uses for which their makers intended them. However, in the age of ecology, the age in which we recognize that everything affects everything, it is no longer possible to remain unaware of the effects of the things we make on our psychic and social lives.

We are living in a situation which has been called "future shock." Future shock, in fact, is culture lag, that is, the failure to notice what is happening in the present.

1. Richard Bach, *Jonathan Livingston Seagull* (New York: The Macmillan Company, 1970), p. 13.
2. Ibid., p. 35.
3. Ibid., p. 83.

Art as Survival
in the Electric Age
(1973)

On April 9, 1973, in a lecture to a large group of undergraduate students at Columbia University in New York City, McLuhan returned to one of his favorite themes, "Art as Survival in the Electric Age."

McLuhan sees the artist as a seer whose role it is to alert us to changes in the environment created by new electric technology. On this particular occasion, he specifies that artistic violence is necessary to jolt people out of dangerous complacency: "One way artists meet the challenge is to involve their public totally in images that are often revolting and irrational. But this technique is to update sensibilities. Any art movement or discovery that has any real core in it enrages people."

McLuhan goes on to talk about popular music in a way that anticipates the development of rap music in our time: "The function of music is to translate the sounds of the environment through language, to humanize the technology of the metropolis by translating those sounds in all their raucous disorder through the rhythms of a great language."

Although McLuhan often astonishes students with his rapid flow of ideas, this time he expresses his own sense of amazement at the invention of instant replay, calling it one of the most

remarkable developments in history, and even going so far as to
name his time "the age of the instant replay."

■ ■ ■ ■ ■

I propose to consider art as a liaison between biology and technology, among other things. And I'm going to venture to read a passage from the work of A. T. Simeons, *Man's Presumptuous Brain*, which indicates a certain gap in our equipment, for which perhaps the artist was intended to close. Dr. Simeons, a biologist, mentions this:

> In man's pre-human ancestors the close and harmonious coordination of cortex and brain-stem was a highly satis-factory means of assuring survival and evolutionary prosperity, as it still is in all wild-living mammals. But when, about half a million years ago, man began very slowly to embark upon the road to cultural advance, an entirely new situation arose. The use of implements and the control of fire introduced artifacts of which the cortex could avail itself for the purposes of living. These artifacts had no relation whatever to the organization of the body and could, therefore, not be integrated into the functioning of the brain-stem.
>
> The brain-stem's great body-regulating centre, the diencephalon, continued to function just as if the artifacts were non-existent. But as the diencephalon is also the organ in which instincts are generated, the earliest humans found themselves faced with a very old problem in a new

garb. Their instinctive behaviour ceased to be appropriate in the new situations which the cortex created by using artifacts. Just as in the pre-mammalian reptiles the new environment in the trees rendered many ancient reflexes pointless, the new artificial environment which man began to build for himself at the dawn of culture made many of his animal reflexes useless.[1]

This enormous gap between man's natural equipment and his technology has gotten bigger and bigger. I suggest that the artist's role is to fill that gap by retuning and modifying the perceptual apparatus that enables us to survive in a rapidly developing environment. Art provides the training and perception, the tuning or updating of the senses during technological advance.

This reminds me of another event, October 4, 1957, the date of *Sputnik*, the date when a man-made environment went around the planet. I think of technologies as extensions of man, of our bodies, of our faculties. I'm not sure what *Sputnik* is an extension of, but it may be an extension of the planet itself. In effect, however, by putting the planet inside a man-made environment, nature ended. Everything that was called nature in preceding centuries ended at that moment, and instead the planet became an art form, an ecologically programmable environment. Ecological dialogue began at that moment and has continued to increase in volume and intensity. I really think that might give certain additional point to this title of today's talk, "Art as Survival in the Electric Age." If *Sputnik* indeed transforms a planet into an art form, then we are living art rather than nature from now on. But instead of making the news or programming the press, we have now to remake the world and program the planet. And since *Sputnik*, we are located in a global theater in which there are no spectators but only actors.

A little boy, when he took his first airplane ride, said, "Daddy, when do we start to get smaller?" I don't know what his daddy said on that occasion but I think he might have paused. It took me quite a while to puzzle that one out. But I ran into a parachutist one day, and he told me that when you jump out of a plane, you are very tiny and you grow bigger and bigger as you approach the earth. There is no *figure/ground* relationship in visual space. The enclosed space of the airplane cabin is static. It's visual. Visual space has very peculiar properties, ones we take for granted even though they belong only to one sense, the visual sense. Visual space is continuous, connected, homogeneous, and static. All the other senses make spaces that are quite different, totally discontinuous, non-homogenous, and dynamic, whether it's the sense of touch, smell, hearing, kinesis, or any sense whatever. There are many spaces formed by our many senses, and all of these spaces are totally lacking in the character of visual space, which is continuous, homogenous, and static. So the moment you jump out of the plane, you jump out of visual space into a live, dynamic space in which you have a *figure/ground* relationship. In the static, enclosed space of the cabin you have no *figure/ground* relationship whatever.

When the ancient Greeks invented visual space by a very peculiar technology called the phonetic alphabet, they fissioned off the visual sense from the other senses, abstracted it from the other senses. And this was done because the phonetic alphabet is phonemic, the only alphabet in the world that is phonemic, not morphemic, that is, the bits of the phonetic alphabet have no meaning. They're phonemes. All other alphabets in the world have meanings. The letters have morphemic meanings. The technology of the phonetic alphabet had this amazing result of fissioning off the visual sense and giving us what is called rational space, Euclidean space, continuous, connected space. And it is useful to know this because today, in the electric age, the simultaneous character of information moving at the speed of light assails from all directions simultaneously as we hear from all directions at once. We do not see from all directions at once, but we hear from above, behind, sides, below, from all directions simultaneously. Acoustic space has a peculiar property; it is a sphere without a center or whose center is everywhere and whose margin is nowhere. This is the space of the vibes. This is the space of electric man. This is the simultaneous space of electric technology. Now this space, needless to say, is scarcely compatible with the static, continuous, connected spaces of the visual, civilized man of the past 2,500 years.

In the early part of the electric age Sam Butler, a biologist, to whom we owe the aphorism, "It is better to have loved and lost than never to have lost at all,"[2] also observed about which came first, the hen or the egg, that naturally, "A hen is only an egg's way of making another egg."[3] This putting the effect before the cause is what we do typically and ordinarily in the electric time.

In 1844, at about the same time that Gould, the mathematician, invented set theory by separating the mathematical operations from mathematical quantities, Edgar Allan Poe, the great innovator in the arts, separated the poetic process from poetry. This was his great breakthrough, and it was of instant effect on the French symbolists and the French poetic activity of the period. Baudelaire translated Poe, or some of him, and took on this idea of simultaneity that if you want to write a poem you have to start with the effect and then look around for the causes. And this became the awareness of acoustic space in which the beginning and the end are at the same time. This is the kind of space and time in which we live now. Einstein was only catching up with Poe in the twentieth century when he invented space-time or relativity theory. The poets and the artists are usually fifty years ahead of the physical scientists in devising models of perception. The job of the artist is to devise means of perceiving that are relevant to the situation in which you exist. This is the gap between biology and technology which Simeons pointed out as a kind of traumatic and dangerous gap indeed.

I'm going to read a short passage from Poe's "A Descent into the Maelström," where he illustrates this principle of survival at work in a situation of a sailor who got caught in the great *ström* and began to study its action as a means to survive:

It was not a new terror that thus affected me, but the dawn of a more exciting *hope*. This hope arose partly from memory, and partly from present observation. I called to mind the great variety of buoyant matter that strewed the coast of Lofoden, having been absorbed and then thrown forth by the Moskoe-strom. By far the greater number of the articles were shattered in the most extraordinary way – so chafed and roughened as to have the appearance of being stuck full of splinters – but then I distinctly recollected that there were *some* of them which were not disfigured at all. Now I could not account for this difference except by supposing that the roughened fragments were the only ones which had been *completely absorbed*. . . . I made, also, three important observations. The first was, that as a general rule, the larger the bodies were, the more rapid their descent – the second, that, between two masses of equal extent, the one spherical, and the other *of any other shape*, the superiority in speed of descent was with the sphere – the third, that, between two masses of equal size, the one cylindrical, and the other of any other shape, the cylinder was absorbed the more slowly. . . . There was one startling circumstance which went a great way in enforcing these observations, and rendering me anxious to turn them to account, and this was that, at every revolution, we passed something like a barrel, or else the yard or the mast of a vessel, whilst many of these things, which had been on our level when I first opened my eyes upon the wonders of the whirlpool, were now high up above us, and seemed to have moved but little from their original station.

I no longer hesitated what to do. I resolved to lash myself securely to the water-cask upon which I now held, to cut it loose from the counter, and to throw myself with it into the water. . . .

The result was precisely what I had hoped it might be. As it is myself who now tell you this tale – as you see that I *did* escape – and as you are already in possession of the mode in which this escape was effected, and must therefore anticipate all that I have further to say – I will bring my story quickly to conclusion.[4]

The point being, he survived by pattern recognition. He perceived the action of the *ström*, that there were certain objects which recurred and survived. He attaches himself to the recurring objects and survives. This is a parable of the artist's role in descending into the dangerous waters. And the *ström* of which Poe spoke in 1850 was nothing compared to the *ström*s in which we are involved at the present moment. Poe hit upon the key to the electric age, programming from effects in order to anticipate causes. The effects come before causes in all situations. The *ground* comes before the *figure* in all situations. So that when any new innovation occurs, people are always able to say, "The time is ripe," meaning the *ground* and the effects have come long before the causes.

A recent article in the July [1972] issue of *Scientific American* on the bicycle pointed out that the bicycle in many respects had paved the way for the motor car literally, because it made smooth roads necessary for the pneumatic car. This is equally true with telephone or with radio or television. The effects come first, and the fact of the effects coming first indicates that the study of environmental action or the action of the *ström* must begin with the effects rather than with a theoretic pursuit of causes. The effects are percepts, and the causes tend to be concepts.

Apropos the structure of the instantaneous or the acoustic environment in which we live electrically, I'm going to read a short passage from Mr. Eliot on the auditory imagination. It is a passage which occurs in his *Use of Poetry and the Use of Criticism*:

> What I call the "auditory imagination" is the feeling for syllable and rhythm, penetrating far below the conscious levels of thought and feeling, invigorating every word; sinking to the most primitive and forgotten, returning to the origin and bringing something back, seeking the beginning and the end. It works through meanings, certainly, or not without meanings in the ordinary sense, and fuses the old and obliterated and the trite, the current, and the new and surprising, the most ancient and the most civilized mentality.[5]

One of the peculiarities of the electric age is that we live simultaneously in all the cultures of the past. All of the past is here and all of the future is here. This is a peculiarity of the instantaneous, the vibes, the acoustic resonance, the acoustic interface which has become the very pattern of our lives. This is a kind of audile-tactile interface which engenders its own patterns of perception, and it's one that early on Yeats drew to our attention. Back in 1900 or earlier, he spoke of the emotion of multitude in a short passage. The emotion of multitude, I think we can discover quite easily, is engendered very much by acoustic resonance and interval, not by connection. He says:

I have been thinking a good deal about plays lately, and I have been wondering why I dislike the clear and logical construction which seems necessary if one is to succeed on the modern stage. It came into my head the other day that this construction, which all the world has learnt from France, has everything of high literature except the emotion of multitude. The Greek drama has got the emotion of multitude from its chorus, which called up famous sorrows, even all the gods and all heroes, to witness, as it were, some well-ordered fable, some action separated but for this from all but itself. The French play delights in the well-ordered fable, but by leaving out the chorus it has created an art where poetry and imagination, always the children of far-off multitudinous things, must of necessity grow less important than the mere will. This is why, I said to myself, French dramatic poetry is so often rhetorical, for what is rhetoric but the will trying to do the work of the imagination? The Shakespearian drama gets the emotion of multitude out of the sub-plot which copies the main plot, much as a shadow upon the wall copies one's body in the firelight. We think of *King Lear* less as the history of one man and his sorrows than as the history of a whole evil time. Lear's shadow is in Gloucester, who also has ungrateful children, and the mind goes on imagining other shadows, shadow beyond shadow, till it has pictured the world. In *Hamlet*, one hardly notices, so subtly is the web woven, that the murder of Hamlet's father and the sorrow of Hamlet are shadowed in the lives of Fortinbras and Ophelia and Laertes, whose fathers, too, have been killed. It is so in all the plays, or in all but all, and very commonly the sub-plot is the main plot working itself out in more ordinary men and women, and so doubly calling up before us the image of multitude.[6]

The sense of universality, which is what he meant by the emotion of multitude, the sense of total involvement in images, is one of the features of the electric time and one of the features of art in the electric age.

Baudelaire drew our attention to a very profound aspect of art and communication in his phrase to the reader of *Les Fleurs du mal*: "*Hypocrite lecteur, – mon semblable, – mon frère!*"[7] The reader puts on the mask of the poem, and the author, the poet, puts on the mask of the reader. It's a reciprocal activity of putting on and putting off in sequence and in simultaneity. This put-on is part of the creation of this sense of order and of multitude and of universality, and this is one of the aspects of art. The updating of this mask of reader and of poet by interplay between them is the interplay of *figure* and *ground* which is one of the needs of people in rapidly changing times.

In *Four Quartets* Eliot mentions the words "slip, slide, perish, decay, / Decay with imprecision, will not stay in place" ("Burnt Norton" V). Eliot has something quite special to say about this mask of language, this mask of putting on the reader. In an essay in which he mentions Mark Twain, in a passage from a volume of his called *To Criticize the Critic*, he says:

It is possible, on the other hand, that the influence of Mark Twain may prove to have been considerable. If so, it is for this reason: that Twain, at least in *Huckleberry Finn*, reveals himself to be one of those writers, of whom there are not a great many in any literature, who have discovered a new way of writing, valid not only for themselves but for others. I should place him, in this respect, even with Dryden and Swift, as one of those rare writers who have brought their language up to date, and in so doing, "purified the dialect of the tribe." In this respect I should

put him above Hawthorne . . . Yet the Salem of Hawthorne
remains a town with a particular tradition, which could not
be anywhere but where it is; whereas the Mississippi of
Mark Twain is not only the river known to those who
voyage on it or live beside it, but the universal river of
human life – more universal, indeed, than the Congo
of Joseph Conrad. For Twain's readers anywhere, the
Mississippi is *the* river. There is in Twain, I think, a great
unconscious depth, which gives to *Huckleberry Finn* this
symbolic value: a symbolism all the more powerful for
being uncalculated and unconscious.[8]

Now this updating of language is really an updating of
sensibility of awareness of perception, and is something which
is the role of the artist to perform. Updating is partly undertaken by popular forms like jazz and rock. And it is curious
that jazz and rock, like much other poetry of our time, depend
upon an oral tradition of speech. Just as the Irish oral tradition
has given us a great deal of contemporary poetry or twentieth-
century poetry, so the only place in the world in which jazz and
rock originate is the deep South. And in the deep South they
have an oral tradition which makes song and dance possible.
But that's *figure*. The hidden *ground* of this activity of popular
music is the sound of the technology of the city. The function
of music is to translate the sounds of the environment through
language, to humanize the technology of the metropolis by
translating those sounds in all their raucous disorder, translating
them through the rhythms of a great language. I suggest that
this is true of any music in any part of the world, but it happens
in our century to be peculiarly jazz and rock forms which are
really a basic type of poetry, updating to oral tradition the sensibilities of modern man.

It does seem paradoxical that the most advanced technology should require a somewhat archaic language. But the

language of industrialized areas has lost its oral quality and oral rhythms and seems to be incapable of translating the new sounds of technology or humanizing the new sounds through that language. I suggest this as a way of looking at or noticing the forms of popular art which have to be participative and have to be related to the new technology; otherwise they would have no public whatever. The response of the young to jazz and rock is no greater than the response of the great poets and painters of our time. The jazz and rock forms are abstract in the sense that they have pulled out the visual connections and the melodic forms. Syncopation in the early phases of jazz simply meant pulling out the visual connections, pulling out the continuity. And in Picasso or any other abstract artist, the technique is simply to pull out the visual connections. That's what abstract means. *Abstractus.* It means the pulling-out. You pull out something. What you pull out in abstract art and in jazz and in symbolism is the connection.

Poe invented not only the symbolist poem by working from the effects before the causes, he also invented the detective story, and the detective story is written by pulling out the connection. The "missing link," and I'm still trying to find out who invented the phrase, the missing link has prompted more participation and scientific endeavor than all the links that were ever made. It's like the detective story. It involves you more deeply. It arouses curiosity more than any possible connection could do. But the strange fact about the missing link or about a discontinuity is that the audience leaps to fill it in, and this is true in jazz and so on. But at any rate, I think it fair to say that jazz and rock have turned English into a world language because you cannot sing jazz or rock in any other language but English. And so the Danes and the Russians and the Chinese and the Japanese have all learned English in order to sing jazz and rock, that is, literally, they do sing it in English whether they know English or not. There's a strange reason, which would take too long to explain, why English alone of all the languages of the world can cope with the musical problem of our time. There's a technical reason, a rather complicated one, that is, complicated to explain.

When Eliot says that he puts Mark Twain equal or even with Dryden and Swift, he indicated that Dryden and Swift had updated the English language in the 1700s by getting rid of rhetoric and getting rid of ornament and using a bare colloquial statement. Now this was against the new *ground* of the Newtonian physics. The *figure* of the English language had to get a new pattern against the hidden *ground* of the new Newtonian physics, and Mark Twain is updating the English language in the electric age by providing a slangy, twangy dialect that relates to oral traditions and brings us back into acoustic space. The electric man lives in acoustic space and has to be related artistically to acoustic space. I'm not saying this is an ideal. I'm simply saying this is actually "where it's at" and "what is happening," and that this is the way artists meet the challenge. They pull out the connections. They involve their public totally in images that are often revolting and irrational and so on. But this technique is to update sensibilities. Violence is a form of identity quest and the updating is often in the form of artistic violence. Any art movement or discovery that has any real core in it enrages people. You can see that we're moving very rapidly to placidity and tranquility in the arts because they've become very monotonously one thing or another, and the time is now ripe for a completely new breakthrough in the arts. I'm not going to take a guess at what it's going to be, but I think it would be worth having since you know that the time is ripe, you know there is going to be a breakthrough. There's going to be a complete change in image in the arts very soon, as there is in technology.

We live in the age of the instant replay, and this is one of the most remarkable developments of any age, since it enables us to have the meaning without the experience. You don't have to watch the game. You can have the meaning of the game minus the experience. The ordinary condition of man is to have the experience without the meaning. This is universal. The replay is a technique not of cognition but recognition. It has changed the nature of sport from top to bottom. It is a form of emotion regulated in tranquility. It reminds one of Mrs. Patrick Campbell's observations about the bliss of the marriage bed after the hurly burly of the chaise longue. Emotion recollected in tranquility was Wordsworth's definition of poetry. The replay is precisely that. The replay, of course, is transforming the football game. The play now has had to be opened up so that people can participate in the process and the technique of the play. Ask any football player and he will tell you that they have to play a totally different game before the cameras than they would ordinarily play. It has to be opened up so that the process of participation in the actual technique of the play can be part of the play. It reminds me a bit of the story about the referee who was moving the ball along fifteen yards for a penalty and one of the players said, "You stink." And so he moved the ball another fifteen yards and turned to the player and said, "How do I smell from here?" That's the replay.

At electric speed you have pattern recognition, and this, incidentally, has implications for many, many things, including education and Watergate or politics. The replay reveals meanings. You know the Watergate thing is a world of meanings minus the experience. We all had the meaning without the experience. Another feature of this same pattern appears in the plight of the motor car at the present time. The motor car has long been the North American's primary mode of privacy. North Americans are the only people in the world (and this is shared by Canadians) who go outside to be alone and inside to be with people. All the rest of the world goes outside to be with people and inside to be alone. And this is a strange story, too.

It draws attention to the circumstances that the most profound things in life are subliminal and unobserved. Europeans don't know that they go out to be with people and inside to be alone. They take it for granted, and we take the reverse for granted. Frank Lloyd Wright was the first architect to recognize this strange American pattern, and to put the bricks in the living room, and the living room on the patio. At the present time the American car has a character unlike that of any other car. It's built for privacy and for going out in to be alone, whereas the European car is, by contrast, like a corset, something very close to the body. It is not intended as something in which you enjoy privacy. And so the motor car is in a strange plight today, the television age. Here's the hidden updating factor. The TV brings the outside inside where people are. It's flipping the whole American way of life and taking the inside outside – Watergate – takes the backroom boys out as show biz and puts the outside world, Vietnam, in the sitting room. Now this has great revolutionary effects on politics. People will not tolerate violence in their homes. They will not have bad news on TV any more. They have had to change the newscasting, cool it, and all the hot stuff is left for the newspapers which is outside, not tolerable inside the home. That's where you go to be friendly and nice and chummy.

So the motor car is being phased out of our lives not by gasoline shortages, but by a complete revolution in our spatial consciousness which is from television. You can tell GM or Ford this some day. They don't know this, and they're not likely to be able to understand it, anyway. But, I mean, without real training in the arts, these people will not survive. This is literally true. Big business cannot survive today without a very highly developed sense of the arts. They're the warning signs. All the warning signals of the new ground are present in the arts long before the hardware boys ever feel them. And so the arts are for survival purposes, and for navigation purposes and as such are indispensable even at the most homely and humble levels.

This particular structure of the North American man who goes outside to be alone and inside to be with people has naturally affected all our literature. You will find that this is the key to American literature whether it's Walt Whitman or Hawthorne or Thoreau. Unlike the European who introduces you to his characters indoors, the North American character is met out-of-doors as an extrovert, a man of the open spaces. He may be a gumshoe or he may be a cowboy, but he's certainly not a man of the salon.

Another peculiar feature of our time is the dropout. The DO is a person who's trying to get in touch by breaking a bind or a hang-up. The problem is related again to the acoustic world of electric institutions. At electric speeds the old organization chart doesn't hold up very well. The educational curriculum chart doesn't hold up very well. The boundaries between subjects and teachers and students don't hold up very well at the speed of light. So the dropout is a person who is attempting to get in touch by introducing a new interval between himself and some bind. The drop-in, on the other hand, is the consultant who is not in quite the same bind. But the wheel and the axle may be useful here as an image of the play that is necessary between *figure* and *ground* in order to be in business or to be in any kind of viable situation. Between the wheel and the axle, there must be a slight interval. It is called play. You may remember an episode that occurred in the newspapers a few months ago, in which some people volunteered to introduce themselves anonymously into mental institutions, only to discover that nobody could tell they were sane. None of the operators of the homes could tell who the sane people were once these people got in. But there was one group of people who had no doubts, and they were the madmen. The mad people pointed at the intruders and said, "They're playing." Now a madman never plays. He is a very serious character, very specialized, very logical. He is granted his very narrow assumptions, and he is absolutely logical. But this lack of play may be the difference between madness and consciousness or sanity. These mad people spotted the sane intruders instantly and said, "They're playing. It's a put-on."

Remember Hamlet, who by way of trying to find out who his friends were, said: I'm going to put on an antic disposition. I'm going to pretend I'm mad. This was a ploy, used by Pirandello in *Henry the Fourth*, in which the Emperor puts on the role of madness in order to ascertain the credibility of his courtiers. Nobody, by the way, can be insincere in the presence of a madman. This, apparently, is a physical and psychological fact, that in the presence of mad people everybody fesses up in some way or other.

The role of the artist in play needs no stress. The artist is a person who tends to use all his faculties simultaneously, and is always at leisure when working hard. The artist is never more at leisure than when he is facing a very tough problem. The use of all the faculties simultaneously creates pleasure. Children always seem to be playing because all of their faculties are simultaneously engaged. They don't have a goal. They do tend to play roles very easily.

One of the peculiarities of innovation is the unexpected retrieval of ancient forms. The electric age, this resonant acoustic time, retrieves, above all, the occult. The occult is by definition the form that is congenial to acoustic man, and he, therefore, appears to be superstitious. I think it was Lévi-Strauss who, in discussing the savage in his book called *The Savage Mind*, says that the savage regards everything as being related to everything, which is a formula for paranoia. The paranoiac is a person who suspects that everything is connected with everything. And this is a very good definition of acoustic man, because acoustically, in terms of the vibes, everything does relate to everything.

One of the functions of the artist that is understood in recent decades is that it is, above all, to prevent us from becoming adjusted to our environments. There's always a danger of becoming a robot, of becoming well-adjusted or conditioned like a man paddling a canoe. A man paddling a canoe may seem very symmetrical and very harmonious in relation to the elements. He is, in fact, a servo mechanism. The better adjusted he is to that paddle, the more a servo mechanism is he. That doesn't mean that it isn't fun. The danger, however, of becoming a servo mechanism of our own environment by adjustment is headed off by the artist who creates violent new images to dislocate our sensibilities. The job of the artist is dislocation of sensibility to prevent us from becoming adjusted to total environments, and to becoming the servant and robots of those environments. That may sound paradoxical. The phrase is from Rimbaud: "*un . . . dérègle-ment de tous les sens.*"[9] The job of the artist is to upset all the senses, and thus to provide new vision and new powers of adjusting to and relating to new situations.

1. Albert T. Simeons, *Man's Presumptuous Brain* (New York: E. P. Dutton, 1962), p. 43.

2. Samuel Butler, *Life and Habit* (London: A. C. Filfield, 1910), p. 13.

3. Samuel Butler, *The Way of All Flesh* (London: Methuen, 1965), p. 297.

4. Edgar Allan Poe, "A Descent into the Maelström," *Poe's Tales of Mystery and Imagination* (New York: Weathervane Books, 1935), pp. 33-35.

5. T. S. Eliot, *The Use of Poetry and the Use of Criticism* (London: Faber & Faber, 1946), pp. 118-19.

6. W. B. Yeats, *Essays and Introductions* (New York: The Macmillan Company, 1961), pp. 215-16.

7. Charles Baudelaire, *Les Fleurs du mal* (Paris: Librairie Jose Corti, 1942), p. 2.

8. T. S. Eliot, *To Criticize the Critic* (New York: Farrar, Straus, & Giroux, 1965), p. 54.

9. Arthur Rimbaud, *Ébauches* (Paris: Mercure de France, 1937), p. 55.

Living at the
Speed of Light
(1974)

On February 25, 1974, McLuhan gave a public lecture to an audience of two thousand people at the University of South Florida in Tampa as part of a College Colloqium Series sponsored by the College of Education. The series focused on future scenarios for education, and McLuhan addressed the shifts that were occurring in media technology and their effects on individual, cultural, and global communications.

McLuhan titles his lecture "Living at the Speed of Light," and describes the movement from the print era to the electronic era in revolutionary terms: "it's helpful to know the origins of the alphabet and of civilization and of rationality in that sense because we have come in the twentieth century to the end of that road. And it's a considerable revolution to have been through 2,500 years of phonetic literacy only to encounter the end of the road. Right now, the people in this room are making the decision whether or not we're going to have any more literacy or any more civilization in the twentieth century, or whether it's going to stop right here."

During the course of this speech, McLuhan explains what he means by his most famous aphorism, "the medium is the message": "It really means a hidden environment of services created by an innovation, and the hidden environment of services is *the thing that changes people. It is the environment that changes people, not the technology."*

■ ■ ■ ■ ■

One of the big flips that's taking place in our time is the changeover from the eye to the ear. Most of us, having grown up in the visual world, are now suddenly confronted with the problems of living in an acoustic world which is, in effect, a world of simultaneous information. The visual world has very peculiar properties, and the acoustic world has quite different properties. The visual world which belongs to the old nineteenth century, and which had been around for quite a while – say from the sixteenth century, anyway – the visual world has the properties of being sort of continuous and connected and homogeneous, all parts more or less alike and static. Things stayed put. If you had a point of view, that stayed put. The acoustic world, which is the electric world of simultaneity, has no continuity, no homogeneity, no connections, and no stasis. Everything is changing. So that's quite a big shift. I mean, to move from one of those worlds to the other is a very big shift. It's the same shift that Alice in Wonderland made when she went through the looking glass. She moved out of the visual world and into the acoustic world when she went through the looking glass.

Now to explain a bit about the implications of this rather large shift. It concerns the whole problem of learning and teaching and social life and politics and entertainment, and I'm going to try to tie it into some of those places. First I will try to make it a little bit more meaningful about how we became visual in the first place.

There is only one part of the world that ever did go visual, and that is the Western Greco-Roman Hellenistic world. About 500 B.C. something happened which made it possible to flip out of the old acoustic world, which was the normal one of the tribal Greek society, the Homeric world. Something happened which flipped them out of the old Homeric world of the bards into this new, rational, philosophically logical, connected, private, individualistic, civilized world. And that thing is called the phonetic alphabet. The origins of the phonetic alphabet are by no means clear at all. All we know is what it did to people. The phonetic alphabet has a very peculiar set of characteristics which are not shared by any other alphabet on this planet. The phonetic alphabet, the one that you all call the ABCs, has a very peculiar structure. It is made up of phonemes, that is, bits that are meaningless. The twenty-six letters of our alphabet have no meaning at all. They're called phonemes because in linguistic terms that means the smallest possible meaningless bit. All the other alphabets in the world, the Hebrew and the Arabic and the Hindu and the Chinese and so on, are morphemic. The bits they have are made of meaning, some meaning, however small.

One of the peculiar things that happened with the phonetic alphabet was that the people who used it underwent a kind of fission. Their sensory life exploded and the visual part of it was cut off from the kinetic, acoustic, and tactile parts. In all the other parts of the world where writing is employed, the visual life has always remained associated with the acoustic life and the tactile life and the kinetic life. The Chinese ideogram is a wonderful instrument of unified sensations. It is so richly unified that most people in our twentieth century have begun to study it very carefully as a corrective to our highly special-ized alphabet. One of the results of the use of the phonetic alphabet was that Euclid could indicate the properties of visual space in his geometry. Visual space, unlike any other of the sensory spaces, is pretty well taken care of by Euclid, who explored most of its dimensions. You've heard of non-Euclidian geometries. Well, in the electric age the non-Euclidian geometries have come back, and Euclid has been put aside. But with the arrival of Euclid and visual space you've got a very strange possibility which Plato seized upon. Plato developed his highly systematized philosophy, even more systematized later by Aristotle, his philosophy of the ideas and the idea of rational control of the passions and of the world of nature.

This Platonic universe of abstract truth and abstract ideas is inconceivable without the phonetic alphabet. This alphabet gave people some very strange habits, too. It filled people with the idea of imperial domination. Western man with his alphabet has always felt it mandatory that he impose it upon all other people. He must spread civilization by spreading literacy in all directions. Now the Romans were the great implementers of this technology. They seized upon this form of writing to codify their laws and to make them uniformly applicable to all men. The idea that civilization, meaning a visually organized set of rules and laws for men in general, the idea that such a thing should be spread to all nations coincided with the rise of Christianity. As far as I know, Christianity has exactly nothing to do with the Greco-Roman idea of civilization. So it is very mysterious that Christianity should have undertaken the job of spreading the Greco-Roman alphabet. At the present time, the church is very doubtful about the matter of spreading Greco-Roman ideas any farther than they've gone, and the Third World doesn't want them. The Third World doesn't want Greco-Roman Hellenistic institutions, the Third World being the non-literate world.

So it's helpful to know the origins of the alphabet and of civilization and of rationality in that sense because we have come in the twentieth century to the end of that road. And it's a considerable revolution to have been through 2,500 years of phonetic literacy only to encounter the end of the road. Right now, the people in this room are making the decision whether or not we're going to have any more literacy or any more civilization in the twentieth century, or whether it's going to stop right here.

One of the strange implications of the phonetic alphabet is private identity. Before phonetic literacy, there had been no private identity. There had only been the tribal group. Homer knows nothing about private identity; Homer's world is that of the acoustic epic, the tribal encyclopedia of memorized wisdom, which Eric Havelock has reported so ably in his *Preface to Plato*. The Homeric epics were part of this acoustic wisdom that preceded literacy and which were phased out by literacy. Homer was wiped out by literacy. Homer had been the educational establishment of the Greeks for centuries. An educated Greek was one who had memorized Homer, who could sing it to his guitar or harp, and perform it in public. He was a gentleman and a free man. Along came the phonetic alphabet, and Plato seized upon it, and said, "Let us abandon Homer and go for rational education." Plato's war on the poets was not a war on poetry, but a war on the oral tradition of education. Today, everyone in this room is being subjected to a new form of oral education. Literacy is still officially the educational establishment but unofficially the oral forms are coming up very fast. This is the meaning of rock. It is a kind of education based upon an oral tradition, an acoustic experience, which is quite strangely remote from literacy. I will be glad to come back to the whole problem of rock and its relation to the modern city and the modern society. It's a very big subject, and it is not very much studied. But rock is not something that is merely stuck onto the entertainment card as an extra item. Rock is a kind of central oral form of education which threatens the whole educational establishment. If Homer was wiped out by literacy, literacy can be wiped out by rock. We're playing the old story backwards, but you should know what the stakes are. The stakes are civilization versus tribalism and groupism, private identity versus corporate identity, and private responsibility versus the group or tribal mandate. This naturally is going to affect our political life, and I'll come onto that shortly.

This is really just an opening theme. I want to mention, by way of explaining my own approach to these matters, that my kind of study in communication is a study of transformation, whereas information theory and all the existing theories of communication that I know of are theories of transportation. All the official theories of communication studied in the schools of North America are theories of how you move data from point A to point B to point C with minimal distortion. That is not what I study at all. Information theory I understand and I use, but information theory is a theory of transportation, and it has nothing to do with the effects which these forms have on you. It's like a railway train concerned with moving goods along a track. The track may be blocked, may be interfered with. The problem in the transportation theory of communication is to get the noise, get the interference off the track and let it go through. Many educators think that the problem in education is just to get the information through, get it past the barrier, the opposition of the young, just to move it and keep it going. I don't have much interest in that theory. My theory or concern is with what these media do to the people who use them. What did writing do to the people who invented it and used it? What do the other media of our time do to the people who use them? Mine is a transformation theory, how people are changed by the instruments they employ. And I wish there were a lot more people in this field that I mention of transformation, but there are extremely few, and I would be embarrassed to mention more than two or three.

One of the peculiar flips that goes with the change from the visual to the acoustic is a change in joke styles. I'm going to tell you a couple of old-fashioned jokes to show you what I mean. A friend of mine went to Kennedy Airport a few months ago to pick up an Irishman who was coming into New York. On the way in from the airport, the Irishman was enjoying the advertising, and he was especially attracted by a sign which read, "Be Younger. Use Ex-Lax." And he said, "How about that. What is Ex-Lax?" His friend said, "We're coming to a drugstore right now and I'm going to get you some." He popped in and brought out a cake of Ex-Lax, which the Irishman proceeded to gobble down *in toto* and with relish. About a half an hour later his friend said, "Are you feeling any younger?" and the Irishman said, "Well, I'm not sure, but I've just done something very foolish." I think he said "childish." Now that's an old-fashioned joke; it's got a story line.

Another one on that pattern concerns a Newfoundland chap who was sitting in an airport waiting for a plane. He was sitting beside another man whom he gradually spoke to. Airports are arranged so that you do not speak to anybody, that is, the chairs are arranged so that you won't be tempted to even notice anybody around you. This is a carefully arranged ploy. Anyway, the man spoke to the Newfoundlander and said, "What do you do?" And the Newfoundlander said, "I'm a rancher. I have forty acres in Newfoundland, and I grow a great variety of things there, and it keeps me very busy." And he said in turn, "What do you do?" And the other chap, who was a Texan, said, "I'm a rancher too." And the Newfoundlander said, "How big is your ranch?" "Well," said the Texan, "if we got in my car about now and drove till sunset, we'd still be on my ranch." And the Newfie said, "Well, you know I had a car like that once." Now that's the old style.

The one-liner joke, which has taken the place of the story line, has no plot at all. It's instantaneous, "easy glum easy glow."

That's the whole thing. That's all the attention span that you're supposed to have anymore. If Nixon had been the captain of the *Titanic*, what would he have said to the passengers? He would have said, "Ladies and gentlemen, we're stopping for ice." These are one-liners. The British Empire is the empire on which the sun never sets because you cannot trust an Englishman in the dark. One-liners are everywhere, and they have taken the place of the old story line.

It's the same way with music. Melody has given place to the new rock forms. Instead of a tune which goes on and on, you simply have the broken and fragmented harmonics and juxtapositions of rhythm, abstract music. Abstract art, abstract music is art in which you pull out the connections. I understand that you're going to have a sculpture by Picasso on this campus. Abstract sculpture or abstract art is an art in which there is no visual component. All you have is the acoustic, tactile, kinetic form. Le Corbusier, the great architect, said that architecture is best appreciated at night in the dark where you can feel the thrust and the forces at work in the building. This is not visual.

Cubism is an art form in which you are given, simultaneously, the underneath, the outside, the top, and the bottom of an object. To have all sides simultaneously is not visual. It is acoustic and tactile. So abstract art is an art in which they have pulled out the visual connections. And that began about 1900. It's about the same time that the physicists pulled out the connections in matter. Quantum mechanics, 1900 – Max Planck pulled out all the connections in matter and gave us quantum theory. Quantum theory is simply physics minus the connections, and it's quite easily understood even by scientists. Don't think they don't have their troubles because one of the problems of Western visual man is that he tries to translate everything into visual terms. It is very difficult for Western man to take things except in a visual, connected, rational mode. Modern physicists report all their findings in Newtonian terms, which are the old-fashioned visual language. One of the peculiarities of modern physics is it still uses the old Newtonian language. Newton was all visual. Everything was classified, connected, continuous. Modern physics has many troubles with the visual problem and the acoustic problem. They don't know whether, for example, to have a particle theory or a wave theory of matter. A particle theory of matter tends to be visual, and a wave theory tends to be kinetic. But modern physics is divided into the different sensory modes of man, and many members of the top physics world are quite unable to understand some of the non-visual aspects of their own field. They're very good at maintaining the general decorum and the conventional respectability of their clan, but in fact they are divided by severe strife within.

Speaking of the flips, there's a story that exists somewhere between the story line and the one-liner, and that is the Norman Mailer story at Berkeley. A few months ago, he was addressing a Women's Lib group and he said to them, "Everybody in this hall who regards me as a male chauvinist pig, hiss," and they all hissed very loudly, and he turned to the chairman and said, "Obedient little bitches, aren't they?" Well, there are two things that joke raises; the new journalism versus the old and Women's Lib. The old journalism used to try to give an objective picture of a situation by giving the pro and the con. Objective journalism meant giving both sides at once. It was strangely assumed that there were two sides to every case. It never occurred to them there might be forty sides or a thousand sides. No, two sides, pro and con, and suddenly this form of journalism disappeared, and the new journalism popped in represented by Truman Capote, Norman Mailer, Tom Wolfe, and many others. The new journalism doesn't give you any side. It just immerses you in the feeling of the whole situation. It just plunges you into the feeling of being at the convention or being at the fire, being somewhere, and it began with that famous phrase, "Something funny happened on the way to the forum." A happening is not a point of view. A happening is all sides at once with everybody involved in it. Mardi Gras is a happening. You cannot have objective journalism about a Mardi Gras. You just have to immerse. Mailer was one of the authors of the new journalism of immersion without any point of view, no objectivity, just subjectivity, and he subheaded his *Armies of the Night* "History as a Novel – The Novel as History." The new journalism quite frankly regards itself as a form of fiction, not objectivity at all.

The new politics is in the same position. The old politics had parties, policies, planks, opposition. The new politics is concerned only with images. The problem in the new politics is to find the right image. So search committees are formed to find the candidates who have the right image. Man-hunting has become a great big business in the military world and the commercial world and the political world. Image-hunting is the new thing. Policies no longer matter because whether your electric light is provided by Republicans or Democrats is rather unimportant compared to the service of light and power and all the other kinds of services that go with our cities. Service environments have taken the place of political policies, or so it seems. I should always add that anything I say is the way it seems at the moment.

The Mailer thing apropos Women's Lib has this rather large implication. Women's Lib is not like the old suffragette thing about votes for women. Women's Lib is not an attempt to find a better, more just set-up for women to be employed in. Women's Lib concerns a tremendous change that's taking place in the entire nature of work. Just as education has undergone strange changes, so has work.

The Japanese Sony plant years ago developed a system whereby all the workers could bring their children to the plant and send them to school. If they were infants, there was daycare, and if they were school-age, they went to school. The Sony plant in Tokyo educated not only the children but educated them at university level, and any of the workers who wanted could also go to university. The plant became itself a kind of playground, and learning and play and work became one thing. Now that isn't too hard to do in Japan, because they are a tribal people and live according to family rules. Nobody ever got fired from a Japanese plant; he is part of the family. Now this tribalism, which they take for granted, is something that they are now trying to get rid of and is something toward which we tend to be moving.

But at present in our own world of work, jobs are giving way to role-playing. Job-holding is giving way to role-playing because at electric speed it is impossible to specialize. This is one of the problems in education. Subjects become very, very dubious as a form of learning. The interdisciplinary takes on more and more meaning. Media study is interdisciplinary study. Isolated subjects in the curriculum have become almost a menace to education. In the same way, the specialized job has become impossible in a big plant or in a big business of any sort. It is more and more necessary to know the overall pattern of the operation.

In Japanese plants like Sony, workers are consulted upon the kinds of innovation, the kinds of products, on their pricing and marketing, on any new developments in the manufacturing process. Everybody in the plant is consulted, not somebody, resulting in total participation on the part of the workers in the whole operation.

The Japanese today are introducing Western literacy into their own culture, and spending $6 billion at the present moment to get rid of their own alphabet and put in our alphabet. Little do they know what is going to happen to them or to us as a result. But the alphabetic man is a very aggressive man and a very specialized man. So the Japanese world is likely to manifest an enormous increase of energy and aggression when they get our alphabet installed. It will also wipe out their whole culture, scrub it right off, that is, their own ideogrammatic forms of writing and culture will be destroyed. Now if China follows the same course – and it appears to be about to do that – then the transformation of the Chinese world will be very rapid, twenty years. They would flip out of their culture, wipe off their whole ancient culture in twenty years and become incredibly aggressive and specialized and goal-oriented, because the specialist man always has a goal. The visual man has a goal in life. The ear man never has a goal; he just wants to do his thing wherever he is. So if the Chinese or the Japanese were to take on our alphabet seriously, they would be in great trouble, and we would too. I don't think they understand what's involved.

Now apropos Women's Lib, the electric world, because it does not favor specialism, does favor women. Men are naturally specialists compared to women. Men are very brittle and unadaptable people compared to women. Through the centuries women have had to adapt to men rather than vice versa.

So specializing, which used to be taken for granted in modern industry, has now become very, very shaky, and role-playing has taken over from job-holding in big business. Role-playing means having several jobs simultaneously or being able to move rapidly from one job to another. A good actor can play many parts. So Women's Lib is really a reply to the new electric conditions of employment in which huge information is available simultaneously to everybody. In the electric world, the simultaneity of information is acoustic in the form that it comes from all directions at once. You hear from all directions at once. Electric information comes from all directions at once, and when information comes from all directions simultaneously, you are living in an acoustic world. It doesn't matter whether you're listening or not, the fact is you're getting this acoustic pattern.

When people become acoustically affected, they no longer have goals. They settle down into role-playing. Some of you may have seen this show called *Upstairs Downstairs* on Sunday nights in which you go down to the servants' quarters. Upstairs is *The Forsyte Saga*, downstairs the servants. In the servants' quarters, people are playing roles. Upstairs, in the Forsyte world of literacy, they are pursuing goals. Downstairs in the servants' quarters in England, the servants had no goals; they just had a role, which was static, but very dramatic, very involving, and very fulfilling.

Now role-playing is a very different thing from goal-seeking, and in the electric time, we are moving very much in that direction. The reason that most of you in this room find it difficult to imagine a goal in life is simply that you're living in an electric world where everything happens at once. It's hard to have a fixed point of view in a world where everything is happening simultaneously. It is hard to have an objective in a world that is changing faster than you can imagine the objective being fulfilled. Women's Lib, therefore, has very deep roots in the new technology, and is not just a matter of votes for women. It means that the work being performed by men today can in many cases be done better by women.

Another strange effect of this electric environment is the total absence of secrecy. What Nixon refers to as the confidentiality of his role and position is no longer feasible. No form of secrecy is possible at electric speed, whether in the patent world, in the fashion world, or in the political world. The pattern sticks out a mile before anybody says anything about it. At electric speed, everything becomes X-ray. Watergate is a nice parable or example of how secrecy was flipped into show business. The backroom boys suddenly found themselves on the stage. Political support for election purposes ceases to be confidential or quiet or secret. There's no way of having any form of secrecy in these matters. With the end of secrecy goes the end of monopolies of knowledge. There can no longer be a monopoly of knowledge in learning, in education, or in power.

I'm not making value judgments. This would seem to many people a very good thing, and it may well be a very good thing. I'm simply specifying the pattern or the form that occurs when you have instant speed of electric information. You cannot have a monopoly of knowledge such as most learned people had a few years ago; you cannot have it under electric conditions. This applies to all professional life as well as to private life.

Ivan Illich has a book called *Deschooling Society*, in which he argues that, since we now live in a world where the information and answers are all outside the school room, let us close the schools. Why spend the child's time inside the school giving him answers that already exist outside? It's a good question, but his suggestion to close the schools is somewhat unnecessary because instead of putting the answers inside the school, it is now possible to put the questions inside.

This might be a good time to mention a little scheme I have for what I call organized ignorance. I've often been puzzled by the fact that the greatest discoveries in the world, when you look back, are perfectly easy. They can be put in a textbook. But the same discovery when you were looking forward at a problem is impossible. Why is knowledge so easy backwards and so hard forwards? Well, it's obvious that this is true because there isn't anything that has been discovered that can't be taught quite easily. Why is it so hard to discover? At first I thought, suppose the cancer experts came to the studio with their problem, set up a model of their experiments and their procedures in studying cancer, and said, "We have got to this point and we cannot get any further." They broadcast that to a million people at once. It is obvious that there'd be one person in a million who would see there was no problem at all. In any problem whatever, one in a million would see no problem. The real problem is how do you reach this guy who sees the absence of the problem.

Now let's ask another question. Why is it that the man, one in a million, says there is no problem? This person is inevitably and naturally untaught, ignorant of all scientific procedures and all science. The scientist has great trouble looking forward past his problem because his knowledge gets in the way. It is only the very ignorant person who can get past that problem because he is not fogged over by knowledge. When you're looking for new answers to new questions, it is knowledge itself that blocks progress. It is knowledge that creates real ignorance, just as wealth creates poverty. Every time a new discovery is made, enormous new areas of ignorance are opened up.

One of the greatest human discoveries, the automatic cybernetic governor on the steam engine, was made by an eight-year-old boy who had the job of pulling the steam cock. Every time the big wheel went around, he pulled the steam cock to let the steam out. He wanted to play marbles; he tied the string to the wheel, and made one of the greatest inventions of all human history. Now the engineers who made the steam engine could not possibly have seen this simple gimmick. Only an ignorant kid who wanted to play marbles could see such things. Now the greatest discoveries in human history are of that kind.

Another strange circumstance attending all discovery and all investigation is this: the effects come before the causes. Without any exception, in every human development, in every discovery, all the effects come before the cause or the discovery itself; so when the discovery is finally made, everybody says, "Well, anybody could have seen that. The time was ripe." About the time somebody discovers the telephone, there are a thousand people who invent the telephone, and then the law courts are filled with suits for generations. Darwin and Wallace discovered evolution at the same time without any personal acquaintance.

At the present time, one of the effects that is heaped up a mile high for which no cause has yet appeared, but a cause will shortly appear, is anti-gravity. We have an enormous amount of anti-gravitational effect and activity – helicopters, airplanes, and astronauts – but we don't have the cause; we just have the effects. Within our lifetimes or your lifetimes, the cause of anti-gravity, a simple gimmick, will present itself, and all things will levitate instantly. The problem will be how to hold things down on the ground. But this is obvious, as obvious as your being there or my being here. The effects are here. The causes will be here shortly.

The bicycle presented all the effects of the motor car just before the motor car. The bicycle paved the way for the motor car, everything, the tires, the chains, and the ball bearings. All the manufacturing problems were solved by the bicycle before the motor car was ever thought of. The roads and the services all arrived first. The motor car arrived last. At the present moment the motor car is on the way out not because of an oil shortage, but because of something quite different. The motor car as a vehicle had an enormous function to perform in American life. It provided the ultimate form of privacy and the means of going outside to be alone. North Americans are the only people in the world who go outside to be alone and inside to be with people. In every other country in the world, including the Eskimo world, people go outside to be with people and inside to be alone.

Why did the Americans ever hit upon this weird reversed pattern? The answer is available. Americans came to this continent to subdue nature fast and furious. They tamed it. They subdued it. They crushed it. They turned it into the enemy. You can read about it in *Moby Dick* or in Hawthorne or in any of our literature. So naturally Americans regard the outside as the enemy, and the inside as the friend, whereas all the other continents in the world regard the outside as the friend and the inside as a place for defence only. All doors are closed in the European house. The European family lives in seclusion and privacy inside. There is no privacy in the American home. That is why you have to get a grant if you want to study so you can leave home. This is a weird pattern, and it's very important to understand it because it isn't over. The motor car provided this superior means of going outside to be alone and, incidentally, going along with it, the great dislike of public transit in America because public transit is where you go outside to be with people, which is very distasteful.

The motor car as the supreme form of privacy has been threatened, in fact superseded by television. Television brings the outside inside, and it takes the inside outside. It pulls the rug out or the highway out from under the car. It deprives the car of its rationale and its meaning. If the car had not lost its real meaning in our lives, there would be no oil price hikes. That is, nobody would even dream of allowing the oil price hikes to occur. The rise in oil prices, of course, is a promotional deal. There's no question. I mean, that's well-known. But it is something that could not have happened if the car had not already been obsolesced.

The car has lost its place in the heart of the people. That doesn't mean it's going to disappear overnight. Not at all. All it means is the effects of the car are disappearing, and privacy and service environments are part of the effects. When I say the medium is the message, I'm saying that the motor car is not a medium. The medium is the highway, the factories, and the oil companies. That is the medium. In other words, the medium of the car is the effects of the car. When you pull the effects away, the meaning of the car is gone. The car as an engineering object has nothing to do with these effects. The car is a *figure* in a *ground* of services. It's when you change the *ground* that you change the car. The car does not operate as the medium, but rather as one of the major effects of the medium. So "the medium is the message" is not a simple remark, and I've always hesitated to explain it. It really means a hidden environment of services created by an innovation, and the hidden environment of services is the thing that changes people. It is the environment that changes people, not the technology.

To come back momentarily to the problem of Illich and the problem of organized ignorance, Illich says we must close our schools because the answers are now outside, and let the kids go back to work and run around the community and get an education. I'm suggesting that the answer is not that, but to put the questions in the classroom, and to start a real dialogue there.

Organized ignorance as a way of bypassing the problem of knowledge as confusion and as block to discovery brings me onto the subject of *Sputnik* and the Laws of the Media. When *Sputnik* went up on October 4, 1957, it put the planet inside a man-made environment for the first time. Spaceship *Earth* has no passengers, only crew. *Sputnik* transformed the planet into Spaceship *Earth* with a program problem. Ecology became the name of the game from the moment of *Sputnik*. Nature ended. The planet became an art form inside a manned capsule, and life will never be the same on this planet again. Nature ended and art took over. Ecology is art.

We now have to confront the need for an ecology of media themselves. It's not just raw materials, but the man-made materials too that now have to be harmonized and resolved in their interaction. Tony Schwartz, in his book called *The*

Responsive Chord, explains this very tricky problem about television as a new environmental medium by saying that the TV image uses the eye as an ear. It's a way of drawing attention to the fact that the TV image has a very different effect on your psychic life than the movie image. Therefore, educationally speaking, TV has very strange consequences and could never be used as a mere transportation device.

The Laws of the Media, which are like the Medes and the Persians, are quite simply this, that every medium exaggerates some function. Spectacles exaggerate or enlarge or enhance the visual function; they obsolesce another function; they retrieve a much older function; and they flip into the opposite form. The simplest form I know to illustrate this principle, which works for all media, whether it's a teaspoon, corset, or motor car, is money. Money increases transactions; it obsolesces barter; it retrieves potlatch or conspicuous waste; and it flips into credit cards, which is not money at all.

Now every medium starts out by exaggerating something that we all have and then finally flipping into the opposite of itself. The motor car flipped into airplane, but first came the bicycle. The Wright brothers were bicycle men. The gyroscopic principle of the bicycle made possible the airplane. The Hula Hoop arrived just before the mini-skirt; the Hula Hoop was a tribal dance which preceded the tribal costume. The effects come first. The cause is later. Now the Laws of the Media I simply mention in passing, but I could spend a long time on them because they are at least a hope that we can reduce this confusion to some sort of order.

What TV Does Best

(1976)

On September 6, 1976, McLuhan appeared on NBC's Tomorrow
Show *with Tom Snyder. As the race for presidency between
Governor Jimmy Carter and President Gerald Ford is in full
swing, Snyder asks McLuhan why Carter seems to come across so
well on television. McLuhan replies that Carter has charisma.
"Charisma means looking like a lot of other people," another of
his aphorisms that turn common sense upside down.*

*When Snyder suddenly says to him, "Has anybody ever asked
you why people find you so hard to understand?" McLuhan tells
him, "It's because I use the right hemisphere of the brain when
they're trying to use the left hemisphere. Simple."*

Snyder: Here's Marshall McLuhan, a man who has been billed as a media guru, and I don't want to use that term, but a man who invented the phrase "the medium is the message," and broke down our electronic stimuli into hot and cool. Have you watched the conventions the last couple of nights?

McLuhan: Well, yes, a little bit.

Snyder: Have they been hot or cool?

McLuhan: Or just apathetic?

Snyder: Or just boring?

McLuhan: "Cool" means really involved and "hot" means standing back, detached. And so what would you say about those conventions?

Snyder: I would say they've been hot, yet if I were on the street, using ordinary jargon or idiomatic expression, I'd say, "Boy, those conventions have been cold," to mean "I haven't been that involved" or "I really haven't been that attracted to what's been happening there." You really have inverted the meaning of the words.

McLuhan: Like the moon shots, they perhaps have overdone it a bit. They have saturated the situation and the thing has sort of backfired.

Snyder: I want to explore a little theory I have about how television audiences can be programmed by what people on television say. If you watch a comedy show where a guest might say to a host, "I really have to tell you a funny story," isn't that word a set-up, and the audience will almost laugh no matter how unfunny the story is?

McLuhan: The funny story has to be based on a grievance so if you can't touch the sore point in the audience, there's no laugh. Most people are waiting for something light, and they are actually going to get stung. The funny story is a very ambiguous thing. It's always based on a grievance.

Snyder: What if someone on television says, "Now Mr. McLuhan, this is really interesting." Won't audiences, no matter what you say, if they aren't paying close attention, and I don't think audiences often pay that close attention, remark to themselves after, "You know, that really was interesting." It's almost as if we're setting them up as we go along.

McLuhan: I think so. They're manipulated. They're part of the pinball machine, if you wish. You ping the ball at them. You hope they'll respond. It's a *figure/ground* situation. You play the *figure* against the *ground*. You rub them together. You hope something will happen. But the *ground* is always hidden. I was watching the pinball machine in your studio, and the hidden *ground* of the pinball machine is the old nostalgia of the 1940s, the days of the Depression. This is the hidden *ground* behind the new interest in the pinball machine, and the pinball machine is now like the movie in the world of TV. The pinball machine is now in a world of electronic simultaneity, and whereas it's an old mechanical machine, one thing at a time, and it belongs to the old hardware days, now it also has a new *ground*. It's the old *figure* in a new *ground*, which makes it an art form. It's now an art form.

Snyder: Before I go too far, what is the world of electronic simultaneity?

McLuhan: All-at-onceness. At the speed of light there is no sequence; everything happens at the same instant. That's acoustic, and everything happens at once. There's no continuity, there's no connection, there's no follow-through, it's just all now. And that, by the way, is the way any sport is. Sports tend to be like that. And in terms of the new lingo of the hemispheres, it's all right-hemisphere. Games are all right-hemisphere because they involve the whole man, and they are all participatory and they are all uncertain. There's no continuity. There's just all surprise, unexpectedness, and total involvement.

Snyder: Is that okay, do you think?

McLuhan: The hemisphere thing?

Snyder: Yes, but I mean the whole thing, all surprise, all spontaneity, no connection, just all at one time. Is that okay for people?

McLuhan: Well, "okay," meaning is it good for people?

Snyder: Yes.

McLuhan: We live in a world where everything is supposed to be one-thing-at-a-time, lineal, connected, logical, and goal-oriented. So, obviously for that left-hemisphere world, this new right-hemisphere dominance is bad. We're now living in a world which pushes the right hemisphere way up because it's an all-at-once world. The right hemisphere is an all-at-once, simultaneous world. So the right hemisphere, by pushing up into dominance, is making the old left-hemisphere world, which is our educational establishment, our political establishment, make it look very foolish. It's just a flip that is taking place.

Snyder: Have you watched enough of Jimmy Carter during all the primaries to figure out why he has been so effective with his presentations on television?

McLuhan: I haven't seen a great deal, but his charisma is very simply identified. He looks like an awful lot of other people. He looks like an all-American boy. He looks like all the American boys that ever were, which is charisma. Charisma means looking like a lot of other people. If you just look like yourself, you have no charisma. So Carter has a lot of built-in charisma of looking like a lot of other guys, very acceptable guys.

Snyder: How helpful would you be to Mr. Carter or whomever the Republicans choose if they were to come to you and say,

"You know, Mr. McLuhan, we'd like to hire you for a specified fee to advise us on a political campaign?"

McLuhan: I could tell them when they were hotting up the image too much and phasing out that charisma. The temptation of any campaign manager is to hot up the image until it alienates everybody, and they don't realize when they're doing it.

Snyder: How do you know when the image is getting too hot?

McLuhan: Specialized, the moment it begins to specialize, it phases out the group.

Snyder: What do you mean "specialize?"

McLuhan: It begins to look more and more like one guy. It begins to look more and more like Jimmy Carter and less and less like the rest of America.

Snyder: Would it be proper to say it's obvious that he's trying to be Jimmy Carter than just Jimmy Carter? Could you pick that point?

McLuhan: Oh, yes. Well, it's just a matter of tuning. It's focusing an image. You can focus in soft focus or hard focus. When it goes hard-focus, that's specialist, and you have lost your audience.

Snyder: Forgive my impertinence, but has anybody asked you why you are sometimes difficult to understand?

McLuhan: Because I use the right hemisphere when they're trying to use the left hemisphere. Simple. You see, ordinarily, people are trained to try to follow you and to connect everything you say with what they last heard. They're not prepared to use their wits. They're only prepared to use the idea they picked up the first time and try to connect it to another idea. So if you're in a situation that is flexible, where you have to use your wits and perceptions, they can't follow you. They have preconceptions that phase them out at once. You see, that's left-hemisphere. I use the right hemisphere a great deal, which is a world of perception, no concepts.

Snyder: Got you, got you, and you don't try to connect things. You just let the right hemisphere take over and let it go.

McLuhan: And watch what's happening. So that's the way the cookie crumbles sort of thing, where you don't know what's going to happen, but you follow the crumble.

Snyder: Could you tell me or any television executive why, for example, Carol Burnett has been able to run for years with a television program, or, as Ed Sullivan did for twenty-three years, and why everything Don Rickles tries doesn't work?

McLuhan: Well, to be honest, I have only seen a bit of Rickles. I've seen more of the other people. They have a mask. They put on a corporate mask or image. They don't use a private face. You remember Ed Sullivan, the great stoneface? He was a corporate mask, like sculpture, and he could identify with a great range of people. Carol Burnett likewise wears a mask, not a private face. Now the mask permits the whole society to identify with that, like a Charlie Chaplin mask, like a Groucho Marx mask. It's not a private face, and so this capacity to put on a huge audience by a mask is the secret of that type of appeal. But it requires a great deal of art and rule of thumb and intuition to fix the mask in a position where it holds a lot of people in focus.

Snyder: Do the people who are able to do that know that they are able to do it, and consciously improve it for greater effect?

McLuhan: They never stop experimenting. They're always tuning and retuning their mask, their image.

Snyder: I wonder if that's true, because again, I go back to the preconception – I'm back on the left hemisphere – that those people who are successful in mass media often can't tell you why they are successful, which would lead me to believe that they don't constantly try to improve their image, or that they

don't step back and look at videotape or look at themselves
and analyze themselves.

McLuhan: They tune. The moment they have their audience
with them they can tell; they can tune their audience, they
can tell when they lose it. Any actor can tell that instantly the
moment he steps on the stage. No two audiences are alike,
and he knows that. It's like a violinist retuning and retuning.
He has to tune himself and retune himself constantly until he
has them on. You put them on.

Snyder: I want to talk for a second about the news on television and about journalism. Eric Sevareid made a talk
defending television news organizations against the charge
that they are biased, and stating that, while many surveys
have been done with viewers to collect their perceptions of
what they feel is biased on television news, there has yet to
be done a survey to measure the bias viewers have built in,
preconceptions, left hemisphere, as they watch a television
program. I wonder what that kind of a survey would turn
up? Is it possible that no matter how objective a report
might be or a piece of documentary film, we all see what we
wish to see and we all hear what we wish to hear?

McLuhan: The user is always the content of any situation,
whether it's driving a car or wearing clothes or watching a
show. The user is the content. However, the news situation
involves strange components. Real hot news that people are
interested in tends to be bad news because that gives them a
thrill of survivor emotion. And the good news, if there is any,
does not give them that survivor feeling. And so the need for
bad news is very strong, especially when there is an awful lot
of good news around in the form of advertising. However,
there is the fact of the top ten. In any news situation there are
only the top ten stories or the top six stories. It's like the disc-
jockey world; there are only certain stories that are –

Snyder: Well, we say it, "Tonight's top headlines."

McLuhan: The top news, and they are a kind of stereotype that is expected by the public. And if there is a sudden new story popped in there, it throws them off. But I don't think the public recognizes there are only a few stories that are tolerated at any one time. You can't play very many stories at any one time in the news, that is, from day to day there has to be a repetition of a certain few stories.

Snyder: You recall the survey that was taken. They showed people a half-hour news broadcast, and an overwhelming majority could not remember one single story they had seen on that half-hour news broadcast. They couldn't recall it.

McLuhan: That is a very good index of involvement. When you're totally involved, you forget completely. It's only when you have detachment that you remember. And that's why one of the tests of a good ad is, if you can't remember it, it's a good ad.

Snyder: You know, you're saying things that seem almost to be the opposite of old established truths. Cool is hot, hot is cool, involvement you don't remember, if you're not involved, you do remember. Why do I have the feeling that I think I remember the things that were most important to me and that I was most involved in?

McLuhan: Can you think of an example? What sort of situation do you refer to?

Snyder: For example, I, uh.

McLuhan: The unforgettable episode, but no, that isn't fair because–

Snyder: I see what you mean. Sitting here, and you ask that question, I say, "Wait a minute. I can't really remember anything of great significance that took place in my life in the past twenty-four hours. Yet at the time there were many things that seemed very important."

McLuhan: I was here during the tornado or the . . .

Snyder: Hurricane.

McLuhan: And I was amazed at the excitement that that generated in everybody, expectancy. And it was covered so thoroughly that it dissipated the storm itself. The coverage actually got rid of the storm.

Snyder: You know, I had the same feeling because we did two hours here on NBC warning people that the sky was going to fall and that they were going to be damned lucky if they survived till morning. The next day it was sunny and warm. There was a little water in the streets, but that's all.

McLuhan: I think that is one of the functions of news, to blow up a storm so big that you can dissipate it by coverage. It's a way of getting rid of the pressure by coverage, that you can actually dissipate a situation by giving it maximal coverage. It's very disappointing from one angle, but it's survival from another.

Snyder: Now don't you get into alarming people?

McLuhan: That's done by rumors, not by coverage, hints, suggestions. But the big coverage merely enables people to get together and enjoy a sort of a group emotion. It's like being at a ball game, a big group emotion. But I do think that that taught me that one of the mysteries of coverage is that it's a way of releasing tension and pressure, and that storm was doomed as soon as it got all that coverage.

Snyder: You're absolutely right, because the event was magnified beyond any proportion. Anticipations were raised that could not possibly be fulfilled.

McLuhan: But the vibrations released to the whole population were quite effective meteorologically.

Snyder: What do you think is the most, I want to use the word *effective* but that's not the right word. I'm talking about television here. What has the greatest impact on the

audience? Is television best when it covers an event like a
space shot or the Olympics or a baseball game? Is it best
when it tries to entertain with movies at night, when it tries
to inform with news programs that have film of things that
have already happened?

McLuhan: The advantage of coverage of sports events is they
are ritualistic. The group gathered there is participating in
a ritual. Now the Olympics were even more a group ritual
than the ordinary competitive event in a ball game or a
single ball game, a single event because they had a corporate meaning. It was not just local. It had sort of a worldwide
meaning. This is itself a ritualistic participation in a large
process. Television fosters and favors a world of corporate
participation in ritualistic programming. That's what I mean
when I say it's a cool medium. It's not a hot medium. A hot
medium can, like a newspaper, cover single events with very
high intensity. TV is not good at covering single events. It
needs a ritual, a rhythm, and a pattern. And that's why a lot
of advertising on TV you see is too hot, too specialized, too
fragmentary. It doesn't have that ritualistic flow. But the
advertisers are aware of this and they're doing a lot to
correct it. But I think that was the great secret of a thing like
the Olympics. People had the feeling of participating as a
group in a great meaningful ritual. And it didn't much
matter who won. That wasn't the point. But I think TV tends
to foster that type of pattern in events. You might say it
tends to foster patterns rather than events.

Snyder: What would happen if you could shut off television for
thirty days in the entire United States of America?

McLuhan: It would be a kind of a hangover effect because it's
a very addictive medium, and you take it away and people
develop all the symptoms of a hangover, very uncomfortable. It was tried, remember, a few years ago; two or three

years ago they actually paid people not to watch TV for a few months.

Snyder: I don't recall that.

McLuhan: It was in the U.K., and they discovered they had all the withdrawal symptoms of drug addicts, very uncomfortable, all the trauma of withdrawal symptoms. TV is a very, very involving medium, and it is a form of inner trip, and so people do miss it.

Snyder: The thought just occurred to me that possibly if you turned off television, there would be a lot of people who would say at the end of the thirty-day period, we will not permit you to turn it back on. Do you think that could happen?

McLuhan: A great many of the teenagers have stopped watching television. They're saturated. Saturation is a possibility. About the possibility of reneging on any future TV, I doubt it. I doubt that, except through saturation. But the TV thing is so demanding and, therefore, so soporific, that it requires an enormous amount of energy to participate in. You don't have that freedom of detachment.

Snyder: How often have we heard people say, "I'm so tired. I've been watching television all evening." It's true. You get involved with these things and you get very, very tired. Do you think we make too much of a big deal about media and about television? After all, we're not curing sick people, we're not feeding hungry people, we're not making the world safe for Christianity or Judaism or democracy. We're just talking about basic television programs.

McLuhan: Yes, well, one of the effects of television is to remove people's private identity. They become corporate peer-group people just by watching it. They lose interest in being private individuals. And so this is one of the hidden and perhaps insidious effects of television. Movies didn't do that. They did not remove people's private identity. It's a very different medium.

Snyder: I'm being told we're running out of time. Thank you, Marshall McLuhan, for being with me this morning. I appreciated meeting you, and although at times I've had difficulty understanding your written word, it was easier listening to you than it is reading you.

TV as a
Debating Medium
(1976)

On September 24, 1976, McLuhan was invited on NBC's Today
Show *with Tom Brokaw and Edwin Newman to talk about
the television debate between Jimmy Carter and Gerald Ford the
night before. In introducing McLuhan, Brokaw makes what was
a common mistake, identifying him as the author of a book called*
The Medium Is the Message. *McLuhan wrote the book* The
Medium Is the Massage, *but "the medium is the message" was
simply the most famous of all his aphorisms.*

*The interview focuses on the outcome of the debate, and
McLuhan makes the controversial point that television is not
effective as a debating medium because of the short attention span
of the audience. He also has great fun with the fact that the debate
has to be halted halfway through due to an electrical malfunction
– with the cameras still running and the two candidates standing
stock-still and silent at their lecterns like a pair of mannequins or,
as McLuhan puts it, two men standing in barrels while their pants
are being pressed.*

Brokaw: Marshall McLuhan is practically a household word in this television-conscious society. He's from the University of Toronto and the author of a book called *The Medium Is the Message*. And he came to New York last night so that he could watch the debates, watch them not so much from a policy point of view, but from a television point of view. First of all, Professor McLuhan, how did you watch the debates, on a number of television sets?

McLuhan: Yes, I saw it on black-and-white and two different kinds of color, the CBS color and the NBC color. They seemed to be quite different on the set I was watching.

Brokaw: It might have been the set.

McLuhan: It might have been the set. On the other hand, the glorious moment was the rebellion of the medium against the bloody message. The medium finally rebelled against the most stupid arrangement of any debate in the history of debating.

Brokaw: Why was it stupid? Not from a political–

McLuhan: From the scripting point of view, the characters who had arranged that debate and scripted every aspect of it had no understanding of TV, and they didn't even know that TV is not a debating medium. And they had arranged it as if it were a newspaper set-up or a radio set-up. They had no awareness of TV. With the breakdown in the technology, the audience finally got into the act.

Newman: Well, Professor McLuhan, if this debate had been arranged by people who in your view knew and understood television, how would it have been done?

McLuhan: Now this would take quite a while to explain, but it would be much closer to what we're doing right here, chatting casually, spontaneously, without a script, and paying attention to what is being said. What those men said last night was merely to hold the audience on the image. It didn't matter at all what was said last night. The image is what mattered, and anything that could hold the attention on the image was all that mattered from the point of view of the arrangers.

Brokaw: You are the proponent that television is a cool medium.

McLuhan: Yes.

Brokaw: Both candidates last night were programmed and costumed and made-up precisely.

McLuhan: Yes, had either candidate dared to present a policy, it would have destroyed his image.

Brokaw: Was one more cool than the other from a purely television point of view?

McLuhan: No, they were both in a sort of state of panic cool. They were terrified of making a false step, and quite rightly so, because –

Brokaw: Why shouldn't they be terrified? Why shouldn't they have all of these conditions on a television debate when they're running for the highest office and they want to control the environment?

McLuhan: With the breakdown in the mechanism, there was the wonderful revelation of all the characters who had scripted the show. They came out into the open like something out of the woodwork and revealed that a stupid show had been put together very carefully and by people who had no idea of what the TV medium is about.

Newman: Would you argue, Professor McLuhan, that the candidates would be better off not to debate than to debate in the manner in which they did?

McLuhan: Of course. They were standing in press-the-pants barrels, looking absolutely like some straitjacketed characters, absolutely the hottest type of medium you could imagine. Everything that the scripters and arrangers had done was hot stuff. They had no idea of what the TV medium is made of.

Brokaw: Perhaps they preferred it that way because there was less risk, however, from a political point of view.

McLuhan: No, it's not the candidates that had anything to do with this, obviously. This was the experts.

Brokaw: The candidates did have a great deal to do with it because all of the arrangements were made after having been cleared by –

McLuhan: By their experts.

Brokaw: By the candidates themselves. They were in on it too.

McLuhan: They couldn't have known that little about what they were doing. But I never saw a more atrocious misuse of the TV medium. When it broke down, it was the thing rebelling against misuse.

Brokaw: You mean you preferred the interviews in the lobby with Rosalynn Carter?

McLuhan: The interval was audience participation and was real television. But what went on in the debates was not television at all.

Newman: Now last night, after all of this happened, the word *amplifier* was being passed around. What went wrong? It was an amplifier. Are you suggesting the amplifier took on certain human qualities?

McLuhan: The vibrations got through to the amplifier and said "This must not continue," an abominable show arranged by a lot of hot experts who –

Newman: The abominable showman.

McLuhan: Yes, the abominable showman took over and blew the amplifier. The vibes really did get through. The medium was the message.

Newman: Professor McLuhan, you understand that although the way you talk about experts, as Tom says, the experts are doing the bidding of their employers. Their employers were President Ford and Governor Carter.

McLuhan: You might as well tell me that that was true of Nixon, who had lots of experts working on his image and they didn't know what they were doing.

Brokaw: They were less conscious of television then than they are now, and what it can do to you.

McLuhan: I wonder. They hadn't a clue as to what the medium was made of.

Newman: Would it be possible that what is going on here is essentially a defensive operation by both men, both of them intent on not making any mistakes, and being plunged into this debate because they can't avoid it? They do everything they can to minimize any risk. They do everything they can to hold the position.

McLuhan: When I was listening to those statistics, I was thinking of the drunk who clung to the lamppost more for support than for illumination. There was no illumination coming out of those statistics, but there was a lot of drunken clutching at lampposts. This was an incredible show of incompetence and misuse of that medium.

Brokaw: But there was an exposition of many of their particular points of view. For instance, I was struck by the pinning down by Elizabeth Drew of the *New Yorker* of the two candidates on their plans for tax reform, balanced budgets, and for social programs.

McLuhan: Why did Ms. Drew and Ganon and Reynolds look so much more impressive than the candidates? This is another medium trick. All those three sounded most authoritative and emphatic, and they knew what they were talking about.

Brokaw: Because it's always easier to ask a question than it is to answer.

McLuhan: No, but I mean the way they delivered themselves was authoritative, emphatic like a good advertisement, whereas the candidates were groping around in the fog.

Newman: Is it possible there is so much at stake for the candidates in an event of this kind that they cannot be at ease? That they cannot be, as you would say, authoritative?

McLuhan: If they were allowed to chat with each other they would soon be at ease. But to stand up in press-the-pants barrels with all sorts of equipment around them and pretend they're having a debate or a dialogue is nonsense.

Brokaw: You're opposed to the ninety-minute format as well. You think that's too long a time?

McLuhan: Well, no television audience can keep its attention more than four or five minutes on anything.

Brokaw: You think the presidential debates ought to consist of two men sitting there for four or five minutes just chatting?

McLuhan: No, but if they were chatting you might be able to hold attention a little longer, but the TV attention span is very short.

Brokaw: What do we have, then, a series of fifteen-minute conversations, one a day?

McLuhan: That sounds reasonable in many ways, but there are other factors. You're assuming that what these people say is important. All that matters is that they hold that audience on their image no matter what they say.

Newman: What image do you think Ford gave out? And what image do you think Carter gave out?

McLuhan: Ford, incidentally, sounds better when you turn off the image and listen to him as if on radio. He's better than Carter. He's a lawyer. Carter doesn't sound as good as Ford with just the soundtrack. Carter looks much better on color than on black-and-white. That is very significant. Ford looks very much better on black-and-white than on color, and this has a lot to do with the constitution, the structure, of those images. Carter is, after all, a corporate man. He comes from a corporate culture. Ford comes from an individualist, fragmented culture, a Northern culture. There's a huge gap between those images.

Brokaw: Carter would probably disagree with you about the corporate structure of –

McLuhan: I'm talking about the South as a corporate territory. His accent is corporate, not private.

Brokaw: What about the impact of his accent on a country as diverse as this one is with its many regionalisms?

McLuhan: Well, obviously it would appeal enormously more to the young generation than to the older one, because jazz and rock come up out of that part of the country, the South.

Newman: When you speak of a corporate area, you're talking about an area that has a degree of cohesion that the other areas don't?

McLuhan: And shares a common accent, a socially shared accent or speech rhythm, which is very potent and very, very cohesive and corporate. It's not a private voice. Ford uses a private voice.

Newman: As you analyze these two candidates, Professor McLuhan, do you think that, given what you've been talking about, one of them has an advantage?

McLuhan: Yes, definitely. Carter has a huge TV advantage if he can keep his foot out of his mouth; just as an image he has a huge TV advantage. No comparison. But it depends on his keeping out of the press. He would do very well to hide for the next six months.

Brokaw: Professor McLuhan, you think that this medium is the future, that this is how society will be shaping its opinions and so on?

McLuhan: Is the present which is also the future.

Brokaw: But why do you continue to print books and write books?

McLuhan: This is an outlet, and you might as well ask why does somebody continue to make chewing gum. It's an outlet for various activities. But I've never been against the book, for heaven's sake. I'm a professor of literature; I teach books from morning till night.

Violence as a
Quest for Identity
(1977)

On December 28, 1977, in what would turn out to be his final television appearance, McLuhan was a guest on TVOntario's public affairs program, The Education of Mike McManus.

Contrary to the common misunderstanding of his term "global village" to mean a place of harmony, McLuhan uses it to describe a tribal world that is savage. "When people get close together, they get more and more savage, impatient with each other . . . the global village is a place of very arduous interfaces and very abrasive situations." In talking about his own country, McLuhan expresses a radical view of Canada's immigration policy. He sees it as a drastic move "to keep people apart and to keep them intact without merging." The cultural mosaic as an idea for immigration is "an amazing strategy of survival. Survival, however, is a legitimate goal in life, especially in a fast-changing world."

In the course of conversation, McLuhan delivers two more fascinating aphorisms: "All forms of violence are quests for identity," and "The literate man is the natural sucker for propaganda," which he explains by saying, "You cannot propagandize a native. You can sell him rum and trinkets, but you cannot sell him ideas."

McManus: Way back in the early fifties, you predicted that the world was becoming a global village. We'd have global consciousness. And I'm wondering now, do think it's happening?

McLuhan: Well, you've heard of Julian Jaynes, *The Bicameral Mind*, the split-up of same and the rise of consciousness.

McManus: So your prediction is correct? We're into it?

McLuhan: No, no, I think now we're playing it backwards. We're going back into the bicameral mind, which is tribal, collective, without any individual consciousness.

McManus: But it seems, Dr. McLuhan, that this tribal world is not friendly.

McLuhan: Oh no, tribal people, one of their main kinds of sport is butchering each other. It's a full-time sport in tribal societies.

McManus: But I had some idea that as we got global and tribal we were going to try to–

McLuhan: The closer you get together, the more you like each other? There's no evidence of that in any situation that we've ever heard of. When people get close together, they get more and more savage, impatient with each other.

McManus: Well, why is it? Because of the nature of man?

McLuhan: His tolerance is tested in those narrow circumstances very much. Village people aren't that much in love with each other. The global village is a place of a very arduous interfaces and very abrasive situations.

McManus: Do you see any pattern of this in, for example, the desire of Quebec to separate?

McLuhan: I should think that they are feeling very abrasive about the English community – the way the American South felt about the Yankee North a hundred years ago.

McManus: But is it a need for space?

McLuhan: No, it's a need for less abrasive encounters, a little more space between the wheel and the axle. When the wheel and the axle get too close together, they lose that playfulness. There's no play left. So they have to have a bit of distancing from each other.

McManus: Is this distancing going to be a pattern right around the world?

McLuhan: Apparently. Separatisms are very frequent all over the globe at the present time. Every country in the world is loaded with regionalistic, nationalistic little groups. Even Belgium has a big separatist movement.

McManus: But in Quebec, for example, you define it as the quest for identity?

McLuhan: Yes, all forms of violence are quests for identity. When you live out on the frontier, you have no identity. You're a nobody. Therefore you get very tough. You have to prove that you are somebody, and so you become very violent. And so identity is always accompanied by violence. This seems paradoxical to you? Ordinary people find the need for violence as they lose their identities. So it's only the threat to people's identity that makes them violent. Terrorists, hijackers, these are people minus identity. They are determined to make it somehow, to get coverage, to get noticed.

McManus: And all this is somehow an effect of the electronic age?

McLuhan: Oh no, but people in all times have been this way. But in our time, when things happen very quickly, there's very little time to adjust to new situations at the speed of light. There's very little time to get accustomed to anything.

McManus: Would, then, the quest for identity of the French Canadians and the kind of inherent violence that you speak of that's concomitant with that not have come so soon without –

McLuhan: Without electric technology, yes, that's true. Things like radio can push people up into a new kind of awareness which makes it very difficult for them to relate to other people. Ireland has shown many responses to this situation in its relations with the North and South of Ireland and in its relations with England. I mention them because everybody tends to know a little bit about that. And it has been irreconcilable until now, anyway, the English representing a highly literate society and the Irish a more oral and much more communal tribal group, and where the tribal feelings are strong, radio sends them up the wall. So radio has sent tribal societies around the globe up the wall with intensity of feeling. One of the big violence-makers of our century has been radio. Hitler was entirely a radio man and a tribal man.

McManus: And what does television do, then, to that tribal man?

McLuhan: Well, I don't think Hitler would have lasted long on TV. Like Senator Joe McCarthy, he would have looked foolish. He was a very hot character, like Nixon. Nixon made a very bad image on television. He was far too hot a character, much better on radio, not bad on the movies, which would take quite hot characters. But Nixon was hopeless on TV.

McManus: The investigations now of the CIA, the FBI, and even our own, God forbid, RCMP, has this anything to do with the electronic age?

McLuhan: Well, yes, because we now have the means to keep everybody under surveillance. No matter what part of the world they're in, we can put them under surveillance. It has become one of the main occupations of mankind, just watching other people and keeping a record of their goings-on. This is the way most businesses are run. Every business has a huge espionage sector. This is called public relations and audience research, and this is around the clock. This has become the main business of mankind, just watching the other guy.

McManus: And invading privacy.

McLuhan: Invading privacy, in fact, just ignoring it. Everybody has become porous. The light and the message go right through us. By the way, at this moment we are on the air, and on the air we do not have any physical body. When you're on the telephone or on radio or on TV, you don't have a physical body. You're just an image on the air. When you don't have a physical body, you are a discarnate being. You have a very different relation to the world around you. And this, I think, has been one of the big effects of the electric age. It has deprived people really of their private identity.

McManus: So that's what this is doing to me?

McLuhan: Yes, everybody tends to merge his identity with other people at the speed of light. It's called being mass man. It began quite a long time ago.

McManus: New technology, you say, is a revolutionizing agent?

McLuhan: Yes, it creates new situations to which people have very little time to adjust. They become alienated from themselves very quickly, and then they seek all sorts of bizarre outlets to establish some sort of identity by put-ons. Show business has become one way of establishing identity by just put-ons, and without the put-on you're a nobody. And so people are learning show business as an ordinary daily way of survival. It's called role-playing. Role-playing has become the normal mode of survival in the business world. Jobs have disappeared, but role-playing has come in on a huge scale, and it's much more flexible than job-holding. Jobs are rather static, repetitive things, whereas role-playing is very flexible. You can play many roles, but you can have only one job at a time.

McManus: Now we've reached the point in time where everyone twenty-four years and under is the TV generation, right?

McLuhan: Yes.

McManus: Do you feel these young people out there under twenty-four have been totally tribalized?

McLuhan: They have lost their sense of direction. They do not have goals. They don't have objectives. And that is putting it mildly.

McManus: You think that's new?

McLuhan: I think that is typical of the twenty-four-years and under, and yes, I think that's new.

McManus: You say, too, that between today's child, who has been raised electronically and who must still live in a literate world, because we are still in a literate world, there is a 2,400-year gap between that boy or girl and his parents?

McLuhan: And his parents, who grew up in a literate society. Well, the alphabet, the phonetic alphabet, the beginnings of Western literacy, came in about 500 B.C., and since then, between then and now, is approximately 2,400, 2,500 years. And we are the first post-literate generation, as it were. We have bypassed the literate world of hardware and the lineal left-hemisphere technology. We have bypassed it by moving once again into the altogether world, the holistic world, of the right-hemisphere people who are the Third World people. So what is happening to our own children is, we are watching them become Third World.

McManus: What does that mean?

McLuhan: It means they feel much more groupy and trendy than they do private or goal-oriented. And so the disc jockeys help this along in a huge way. By the way, one of the big parts of the loss of identity is nostalgia. And so there are revivals on all hands, in every phase of life today, revivals of clothing, of dances, of music, of shows, of everything. We live by the revival. It tells us who we are or were.

McManus: Now these children that are more groupy and less private, are they also more passionate or more violent?

McLuhan: I think the sheer dislocation of their lives has put them through a very violent course indeed. They have been ripped off.

McManus: They're kind of rudderless.

McLuhan: They don't have goals because at the speed of light, what is a goal? You're already there. You name it, and you're there.

McManus: The violence of the media, you say, itself invades those not prepared for it. It's not the content that's primary. It's this invasion of privacy that people are not prepared for that is destructive.

McLuhan: We recently had this trial of the young man who appealed to Kojak as his alibi for murder. This is a pathetic thing, because nobody ever mistook fictional entertainment violence for reality. It's impossible. Only people who are leading a merely drugged fantasy life can do that, and there is the strange factor that television is quite a potent drug. It is addictive. It is an inner trip, and it is a tranquilizer. And recently the *Detroit Free Press* offered $500 to anybody who would stop watching for a few days, and they didn't get many takers. But those who did take up the thing dropped out in a few days. They couldn't bear it.

McManus: Do you feel that the fact that you and I have enjoyed the rewards of literacy, that we are more protected against television than a child?

McLuhan: Yes, I think you get a certain immunity just as you get a certain immunity from booze by literacy. The literate man can carry his liquor, the tribal man cannot. That's why in the Moslem world or in the Native world booze is impossible. It's the demon rum. However, literacy also, though, makes us very accessible to ideas and propaganda. The literate man is the natural sucker for propaganda. You cannot propagandize a native. You can sell him rum and trinkets, but you cannot sell him ideas. Therefore, propaganda is our

Achilles heel. It's our weak point. We will buy anything if it's got a good hard sell tied to it. And so propaganda is the great big soft spot in the makeup of the literate man.

McManus: Electronic people, you say, lose their religion very easily?

McLuhan: Well, their attention span is very weak, as you know. We've invented the one-liner in place of the joke because people can't wait around to hear you tell a joke. It takes too long. "As for critics," said Sam Goldwyn, "don't even ignore them." That's a one-liner. That's all we have time for. Attention span gets very weak at the speed of light, and that goes along with a very weak identity.

McManus: And religion, which involves ideas, requires a little more time?

McLuhan: Religion is a form of indoctrination, which requires a considerable amount of literacy. You cannot get religion into people minus literacy. And as literacy weakens, people lose their religious affiliations.

McManus: What now, briefly, is this thing called media ecology?

McLuhan: It means arranging various media to help each other so they won't cancel each other out, to buttress one medium with another. You might say, for example, that radio is a bigger help to literacy than television, but television might be a very wonderful aid to teaching languages. And so you can do some things on some media that you cannot do on others. And, therefore, if you watch the whole field, you can prevent this waste that comes by one canceling the other out.

McManus: Dr. McLuhan, you have admitted in the past that you hate to see the upheaval that our world is in, true?

McLuhan: I wouldn't say "hate" to see it. It is a very confusing kind of world in which you have no time to get adjusted to anything or acquainted with anybody. You know, we live in a world where you meet many, many people per day for the one and only time in your life or their lives.

McManus: But I'm thinking more about the fact that you were born in Edmonton, mastered in English. You were an English graduate who studied largely at Cambridge. English was largely your world, and literature has been under attack by these new electronic media, especially television. Has this been hard on you?

McLuhan: I don't think so, because there's always the challenge of meeting the opposition head-on.

McManus: But you wouldn't like to see the literate world disintegrate?

McLuhan: By no means. My values are strongly centered in literacy, which I teach day and night.

McManus: Do you think it will survive?

McLuhan: I imagine so, I think so.

McManus: It said, too, you felt hostile to modern life, that you loathe machinery, and you hate big cities?

McLuhan: Well, are they talking about a period when I wrote *The Mechanical Bride*? It's a little while ago, all right. I haven't had much time to indulge those feelings. I've been too busy to develop those hostilities, and I've had wonderful luck in meeting fascinating people and having wonderful students year in and year out, and so the amount of satisfaction is huge, and it would be a very selfish thing to blame anybody for anything else.

McManus: It also said you've never been lonely for a moment in your life?

McLuhan: That's true, I've never had that experience.

McManus: Describes your conversion to Roman Catholicism in 1938 as a long pilgrimage and a solitary one, done entirely by reading.

McLuhan: That's, I think, true, except again I had luck. I met people too, but it was mainly a literary activity.

McManus: A book by Gilbert Keith Chesterton, *What's Wrong with the World?*

McLuhan: Handed to me on a Winnipeg street by Tom Easterbrook. He said to me, "I hated this book. I think you'll like it."

McManus: I read some Chesterton in my younger days, and I've often wondered if you'd be satisfied with your contribution, on a much more global scale than Chesterton's, but as being something similar to his. He was always taking the accepted and turning it upside down and inside out.

McLuhan: Having a good look from many sides. He was cubist, you see. A paradox is a form of cubism in which you look at the same situation simultaneously from different directions.

McManus: So there are some parallels?

McLuhan: Well, sure, the habit of discontinuous and multi-leveled perception, but it goes partly also with in my interest in Joyce, Pound, Eliot, because they are also multi-faceted people and very right-hemisphere people.

McManus: And Harold Innis?

McLuhan: I was very lucky to encounter him. It was through *The Mechanical Bride* that I met him. When I heard that he had put it on his reading list, I was fascinated to find out what sort of an academic would put a book like *The Mechanical Bride* on a reading list. So I went around and met him and we became acquainted for the few years of his life that remained. He only had about three years to live at that time. But Innis, I think, is the only man since the beginnings of literacy 2,400 years ago who ever studied the effects of technology, and I think that is an amazing thing in view of the numbers of great minds that had this opportunity. He's the only human being that ever studied the effects of literacy on the people who were literate or the effects of anything on anybody. This, as I say, is a unique thing in Innis's case. Aristotle and Plato never studied the effects of anything on anybody.

McManus: Would you list the The *Mechanical Bride, The Gutenberg Galaxy,* and *Understanding Media* as the three monuments of yours as far as books are concerned?

McLuhan: I have a new book called *The Laws of the Media* which I hope will be much more attractive, but I'm working on new things all the time.

McManus: And still looking for pattern recognition?

McLuhan: It's one of the big excitements of life. It's a sort of detective activity, you know. I do a lot of sleuthing.

McManus: On being a Canadian, I think Northrop Frye says that one of the things Canada gives you is a chance to be an observer.

McLuhan: Yes, because you're not too deeply involved in other people's problems, and our own problems are relatively small compared to other people's problems. And so you can be an Ann Landers to the world.

McManus: What about the multicultural mosaic?

McLuhan: That is an amazing ploy to preserve the cultural identity of Quebec and other minorities.

McManus: Why do you call it a ploy?

McLuhan: It's sort of an official ending of melting pot. The Québécois are terrified of being merged in the American culture. I think it's as simple as that, and I think they're right. They're absolutely vulnerable. We're all vulnerable to the Americans, and they're a very attractive and wonderful people, and I think we could easily become merged in their lives, as we intend to be anyway.

McManus: Does that mean that in order to lessen their fears, we attempt to paralyze different immigrant groups coming in at their state of arrival?

McLuhan: And keep their cultures intact and separate. That is the meaning of the multicultural mosaic. The mosaic is static. It isn't in a state of constant interplay. It's static, and that's exactly the way the French want to remain. They want to remain just the way they are, and so it's not that easy, and so, as I say, this is an amazing ploy developed to make this possible. I don't know if it will work, but I certainly don't wish them ill on this maneuver. It is a kind of media ecology, you see. It's a way of using our available resources in communication to keep people apart and to keep them intact without merging. So I think it's a drastic move. I never heard of it occurring in any other country, did you? I never heard of a multicultural mosaic as an idea for immigration. It's an amazing strategy of survival. Survival, however, is a legitimate goal in life, especially in a fast-changing world.

McManus: And you've felt this yourself?

McLuhan: Oh yes. I have an essay coming out in a Harvard book on Canada called "Canada: The Borderline Case." It is in a book called *The Canadian Imagination*, and the theme has to do with the strange effects of being on so many borderlines in Canada. We have so many cultural border-lines in every direction that it is very confusing to the idea of private identity or even group identity. But it is very enriching, too, because people on frontiers have a very rich life of interplay with other people, other cultures.

McManus: I know you don't like to make predictions.

McLuhan: I make them all the time, but I make absolutely sure they've already happened.

McManus: You don't like being called a prophet, but, of course, I think in the biblical sense there was an understanding of a prophet as someone who is not just talking about the future, but someone who is telling you what is happening now. Would you like to tell us whether the country is going to stay together or not, and thereby be telling us what's happening to us now?

McLuhan: Well, I don't know I'd call that a prophecy or not. There is a sense in which the separatism occurred long ago. But there is a hardware sense in which it is still intact. The country is still intact in a hardware sense, legal sense. It is the hardware that is in danger under electric conditions. The hardware world tends to move into software form at the speed of light.

McManus: You're losing me again, Dr. McLuhan. In fifteen seconds, I have one question for you. How much television do you watch?

McLuhan: Whenever I get a chance. Not too often. I missed *Rigoletto* last night, and I was very disappointed.

McManus: But you don't watch it that often?

McLuhan: No, I don't have that many opportunities.

Man and Media

(1979)

Even though McLuhan always insisted that he was not trying to create a self-contained body of theory, near the end of his life he did leave some clues as to what he intended as the culmination, the ultimate synthesis, of his theory. This came in spring 1979, in his last taped lecture, titled "Man and Media," which he delivered at York University in Toronto.

McLuhan had begun to look at all human artifacts, from the earliest tools to the electric media, including computers, as extensions of the human body and the human nervous system – and as components of a human evolution that Darwin could never have imagined. "Man's technology," he says, "is the most human thing about him," but he hastens to warn that we are completely unequipped to cope with its destructive consequences. To counteract this problem, McLuhan offers a new survival approach to understanding the effects of any new technology, which he calls his "Laws of the Media": the "Laws of the Media are observations on the operation and effects of human artifacts on man and society."

When I was in Barcelona a few weeks ago, I had a translator who was very good, and as I listened to him I noticed something that might be of some relevance. You cannot translate jokes into another language. Now the reason, I think, is that a joke really requires a hidden *ground* of grievance, for which the joke is only a *figure* sitting out front. You remember the streakers? They had a kind of grievance and now it's called "just a passing fanny." But the grievance behind the streakers' performance can't be translated. You cannot verbalize the kinds of grievances that prompted the streakers to go into action. Even the simplest jokes cannot be provided with a ground that makes them plausible.

So I had in mind to tell my Spanish audience a few of our Newfie stories as a way of introducing them to Canada. I thought of stories like the Newfie who went into the bank to cash a check. When asked for identification, he produced a pocket mirror, looked in it, and said, "That's me, all right."

And there's the one about the Newfie who was asked by a sociologist, "Do you have any brothers or sisters?" And he said, "Yes, I have a brother at Harvard." "Oh! What's he studying at Harvard?" "He's not studying at Harvard. They're studying him." Or these: On top of the Newfie ladder there is written on the rung, "Stop here." On the bottom of every Newfie beer bottle, it is written, "Open the other end." How would a Newfie have handled Watergate? "Same way." And so on. Now why are these Newfie jokes? What is this grievance we have against Newfies? If we didn't have a grievance, there would not be any stories.

The Newfie stories happen to be our kind of bundle at the moment, but the same is true in any other language. These jokes do, or try to, cross borders and move from one area to another. Jokes are themselves a very important form of communication, and they reflect certain grievances and irritations which everyone feels. And yet, very little notice is given to them.

In Paris a few months ago the story was going around about the man who was feeling alienated and alone in the world. He walked out and jumped off the top of his high-rise, and as he was passing the floor of his apartment he heard the telephone ringing, the telephone reminding him that there was someone in the world beside himself. Now that's a sick joke. Sick jokes are a new form of grievance and just why they should have their vogue is worth looking into.

In Moscow they tell the story about the attempts to set up an American-style nightclub. This nightclub flopped, and a committee was formed to examine the situation. Questions were asked. "How about the food?" And the managers explained, "We had French chefs and cuisine and the best wine at very reasonable prices." "How about the decor, the layout?" "Well, we had Italian designers with Hollywood consultants." "And how about the girls?" "They were absolutely tops, every one of them party members since 1917." So you can see where the grievance is in Russia.

You cannot have a joke without a grievance. The fast pace in our world has led to the development of the one-liner, the abbreviated joke. The one-liner is for people of very short attention span, who won't stay around long enough for you to tell them a story. You have to work fast. You flip in with a single gag – in one ear and out another. There is the one-liner about Zeus, who says to his fellow god, Narcissus, "Watch yourself." Another one: "You cannot see the writing on the wall till your back is up against it." Or, "Politics is shooting from the lip." The list is simply endless.

These responses to grievances lead to much stronger reactions or statements, such as horror movies, vampire movies, and our general cult of horror. This is a response to situations of the media which people feel are involving them, embedding them. *The Exorcist* is an account of how it feels to live in the electric age, how it feels to be completely taken over by alien forces and hidden powers. The viewer feels he has been obsessed or possessed. And, of course, there is the one-liner, "If you don't pay your exorcist, you will be repossessed."

However, the dropout is the figure of our times. He is the person who is trying to get in touch. When you get uptight you have to let go in order to get back in touch. "To get in touch" is a strange phrase. When a wheel and an axle are playing along together, as long as there is a nice interval between wheel and axle, they are in touch. When the interval gets too big or too small, they lose touch, the wheel is either uptight, or seized up, or else falls apart. Keeping in touch requires this interplay, this interface, which is a kind of interval of resonance. Touch is actually not connection but interval. When you touch an object there is a little space between yourself and the object, a space which resonates. This is play, and without play there cannot be any creative activity in any field at all.

This leads to the theme of violence as a response to situations in which you feel you have lost your identity, situations in which you have been ripped off by too-rapid changes, where you have suddenly been flipped from one situation to another without warning and you suddenly are minus your identity. You don't know who you are. You don't know where you are. This leads to a response of violence.

Incidentally, in a public situation such as an ordinary football game all persons present are nobodies. Even if you were sitting there with Charlie Chaplin or the Duke of Edinburgh, they would be nobodies and you would be a nobody. Anybody at a ball game is a nobody. Now what is the compensation that people expect for becoming nobodies? They expect violence. And so sport provides a systematized,

organized form of violence which compensates the partici-
pants and the audience for being nobodies. The nobody
in real life is a person who becomes quite intractable, quite
violent as a way of rediscovering "Who am I" and "How do I
re-establish my image, myself in this world." Our popular
forms of entertainment tend toward that direction. Whether it's
Westerns or whodunits or horror movies or just about any
other kind of movie, these are movies in which people are
questing for their identity. You may remember an episode in the
novel *A Passage to India* by E. M. Forster, in which Adela
Quested encounters a strange figure in the Marabar Caves. It is
a moment for her of absolute horror. This encounter of a
highly civilized person with an unknown force is a kind of
interplay between the merely visual and the merely familiar,
and the hidden, echoing, resonating, ghostlike world.

In our time, one of the strange things that's happening is
not unrelated to that. Having come out of a very visual age
with organized points of view, positions, jobs, attitudes, we
are suddenly confronted with an instantaneous-simultaneous
world in which there are no familiar boundaries. Moving from
the old hardware world of the nineteenth century and the
industrial or First World technology – the First World technology of the familiar productive and industrial type – into a
world of instant information and design and pattern is a flip
from a visual world to an acoustic world. The main aspect of
our simultaneous-instantaneous time is essentially acoustic, not
visual, since no point-of-view is possible to the ear.

To the world of the simultaneous and the instantaneous
there is no sequence. There is no logic. There is only the simultaneous-instantaneous burst breakthrough. I was once in a
plane when it was struck by lightning. There was suddenly
a bang and a flash and the stewardess casually said, "We have
just been hit by lightning." This had never happened before to
me, but apparently was a common occurrence to her. To a

person who had not a clue as to what had happened it would be rather startling. But this very world of the instantaneous, the simultaneous, is the world in which we live.

What is called the generation gap represents a division between people who grew up in the visual era of the First World with its industrial complex – a world of jobs, of points of view, of policies, and attitudes – and the children of those people, children who were brought up in the acoustic simultaneous-instantaneous world of television. This had happened in a smaller degree in the 1920s with radio. The radio generation was somewhat alienated, dislocated. It was called the "lost generation" by Gertrude Stein. However, the radio generation remained relatively intact compared to the TV generation. Perhaps we should look further into that – the generation gap between parents, who grew up in the old First World, and their children, who grew up in the Fourth World.

The Fourth World is the electric world that goes around the First, Second, and Third worlds. The First World is the industrial world of the nineteenth century. The Second World is Russian socialism. The Third World is the rest of the world, where industrial institutions have yet to establish themselves, and the Fourth World is a world that goes around all of them. The Fourth World is ours. It is the electric world, the computer world, the instantaneous communication world. The Fourth World can come to Africa before the First or Second worlds. Radio came to Africa and began to penetrate African institutions and psyches a long time ago. Radio went to China and India long before anything else from the West. The coming of the Fourth World, the electric instantaneous world, has been completely ignored by the journalists and by the Marxists. Marx, by the way, was a nineteenth-century man, a hardware man of the First World only, who knew nothing about electricity, nothing about the instantaneous. He could not possibly have known what might happen in the Fourth World, an instantaneous world of electric information. His entire thought was based upon production and distribution of product. His conviction was that if everybody could have enough of everything, problems would disappear. It never occurred to him that perhaps the most important commodity in the twentieth century would be information and not hardware products. Information is not only our biggest business, but has become education itself.

But I have digressed slightly. My topic is "Man and Media," a topic which relates to an aspect of media on which I have been working a good deal lately. The preface to a new book of mine begins, "All of man's artifacts, of language, of laws, of ideas, hypotheses, tools, clothing, computers – all of these are extensions of our physical bodies." This power to extend ourselves was used as a theme by Hans Hass in his book *The Human Animal.* In it he considers this human ability to create additional organs as "an enormity from the evolutionary standpoint – an advance laden with unfathomable consequences." My own *Laws of the Media* are observations on the operation and effects of human artifacts on man and society, or, as Hans Hass notes, a human artifact is "not merely an implement for working upon something, but an extension of our body effected by the artificial addition of organs; an advance to which, to a greater or lesser degree, we owe our civilization."[1] Hass considers our bodily extensions as having these advantages: (a) they have no need of constant nourishment, thus they save energy; (b) they can be discarded or stored rather than carried around, a further saving of energy; (c) they are exchangeable, enabling man to specialize, to play many roles, so that when carrying a spear, he can be a hunter or when using a paddle he can move across the sea. All of these instruments can be shared communally. They can be made in any community by specialists, giving rise to handcraft skills.[2]

Something overlooked by Hans Hass was the absence of biological or psychological means of coping with the effects of our own technical ingenuity in creating new organs. The problem is clearly indicated by Albert Simeons in *Man's Presumptuous Brain*, in which he says that

> when, about a half million years ago, man began very slowly to embark upon the road to cultural advance, an entirely new situation arose. The use of implements and the control of fire introduced artifacts of which the cortex could avail itself for purposes of living. These artifacts had no relationship whatever to the organization of the body and could, therefore, not be integrated into the functioning of the brain-stem.
>
> The brain-stem's great body-regulating centre, the diencephalon, continued to function just as if the artifacts were non-existent. But as the diencephalon is also the organ in which instincts are generated, the earliest humans found themselves faced with a very old problem in a new garb. Their instinctive behaviour ceased to be appropriate in the new situations which the cortex created by using artifacts. Just as in the pre-mammalian reptiles the new environment in the trees rendered many ancient reflexes pointless, the new artificial environment which man began to build for himself at the dawn of culture made many of his animal reflexes useless.[3]

What Simeons is saying is that our natural responses to media and to technology are irrelevant, that we cannot trust our instincts or our natural physical responses to new things. They will destroy us. How are we to bypass or offset the merely physical response to new technology and new environments created by new technology?

This problem has not been raised by anybody, even though we have to live with it every day. Edgar Allan Poe's story "A Descent into the Maelström" had tremendous influence on the nineteenth-century poets and symbolists like Baudelaire, Flaubert, and others. In this story, Poe imagines the situation in which a sailor, who has gone out on a fishing expedition, finds himself caught in a huge maelstrom or whirlpool. He sees that his boat will be sucked down into this thing. He begins to study the action of the *ström*, and observes that some things disappear and some things reappear. By studying those things that reappear and attaching himself to one of them, he saves himself. Pattern recognition in the midst of a huge, over-whelming, destructive force is the way out of the maelstrom. The huge vortices of energy created by our media present us with similar possibilities of evasion of consequences of destruc-tion. By studying the patterns of the effects of this huge vortex of energy in which we are involved, it may be possible to program a strategy of evasion and survival.

Survival cannot be trusted to natural response or natural instinct since the brain stem is not provided with any means of responding to man-made environments. Our diencephalon, our huge evolutionary structure of nerves and brain stem, evolved over long periods of time and had ended its development long before the first technology. Long before fire or clothing, this brain stem had completed its programming. And so with the coming of fire and clothing and weapons, the brain stem was unable to respond relevantly to any of these artifacts. The artist's insights or perceptions seem to have been given to mankind as a providential means of bridging the gap between evolution and technology. The artist is able to program, or reprogram, the sensory life in a manner which gives us a navigational chart to get out of the maelstrom created by our own ingenuity. The role of the artist in regard to man and the media is simply survival.

There is a passage in Anthony Storr's *Human Aggression* in which he observes that

It is obviously true that most bomber pilots are no better and no worse than other men. The majority of them given a can of petrol and told to pour it over a child of three and ignite it, would probably disobey the order. Yet, put a decent man in an aeroplane a few hundred feet above a village, and he will, without compunction, drop high explosives and napalm and inflict appalling pain and injury on men, women, and children. The distance between him and the people he is bombing makes them into an impersonal target, no longer human beings like himself with whom he can identify.[4]

This is a characteristic situation. That bomber pilot is very much like the person introducing any new technology using ordinary human business resources and existing institutional means. None of these people ever considers what will be the impact or the effect of what they do when they pull that trigger. Quite apart from the use of weaponry at a distance, there are the effects of changes in man himself which result from using his own devices to create environments of service. Any new service environment such as that created by railways or motor cars or telegraph or radio, deeply modifies the very nature and image of the people who use them. Radical changes of identity happening in very sudden, brief intervals of time have proved more deadly and destructive to human values than were wars fought with hardware weapons.

In the electric age the alteration of human identity by new service environments of information have left whole populations without personal or community values to a degree that far exceeds the effects of food and fuel and energy shortages. I am suggesting that the Club of Rome is really talking to the old nineteenth-century situation of quantity and hardware and ignoring completely the effect of software information on the human psyche. The rip-off of human psychic resources by new media may far exceed the dangers involved in energy shortages from hardware.

I am going to introduce a new survival approach in my forthcoming book on *Laws of the Media* and I hope that readers will offer many improvements to this method. In the meantime, I suggest that it is possible to notice, to understand, the effects of any technology, whether new or old, by applying these four questions to the situation:

1. What does the technology amplify, enhance, or enlarge?
2. What does it obsolesce?
3. What does it retrieve or bring back from a distant past? (probably something that was scrapped earlier).
4. What does it flip or suddenly reverse into when pushed to its limits?

I will give you a few examples of this pattern, these four phases or stages in the development of any artifact whatever. I have in front of me, in isolation from other things, a camera. By its snapshotting quality it enhances aggression and private power over people. It obsolesces privacy. It retrieves the past as present; it brings back the big-game hunter. Bringing him home alive means bringing people home alive: photographic journalism is very big-game hunting. It flips into the public domain.

The zipper, the homely zipper. It amplifies the grip, the clasp. It obsolesces the button, the snap. It retrieves long-flowing robes, easy to manage. It reverses into Velcro drape, no clasps, no buttons, no zipper, no closure at all.

The clock amplifies work. Until the clock was invented, what we call work was almost impossible to organize. It obsolesces leisure. It retrieves history as art form by fixed chronology – immeasurable, sequential chronology capable of visual time as measured by the clock. It reverses when pushed all the way into the eternal present, a nowness.

Instant replay enhances awareness of the cognitive process. It obsolesces the representational, the chronological, in that it doesn't matter in what sequence the events occurred. It retrieves meaning. You can have the meaning in instant replay without the experience. This is a rather startling aspect of the instant replay; you can have the meaning, the structure, minus the experience of the event. And it flips into corporate pattern recognition, which is easily associated with tradition. The instant replay is, perhaps, the most remarkable development of our time, and one of the most profound and metaphysical.

Electric media in general amplify information, range, and scope, pushing information into a service environment by simultaneity. Electric media obsolesce the visual, the connected, the logical, the rational. They retrieve the subliminal, audile, tactile, dialogue, involvement. They reverse finally all hardware into software. The motor car is only worth a few bucks in hardware terms; it is worth many millions in software terms of design. And consider electric speeds. At electric speeds the sender is sent. The sender goes on the air and is instantaneously everywhere without a body. Electricity creates the angelic or discarnate being of electronic man, who has no body. When you are on the telephone, you are in New York or here, simultaneously, and so is the person you are speaking to, minus the body. The implication of discarnate, disembodied existence in an information world is one for which our educational system has not quite prepared us.

This pattern of four aspects of change – enhancing, obsolescing, retrieving, and flipping – happens to be the pattern of a metaphor. All metaphors have these four aspects. All metaphors are *figure/ground* in ratio to *figure/ground*. They are not connected. They are in ratio. Metaphors are in. All technologies have these four aspects. I was gradually forced to conclude that all human extensions are utterings or outerings of our own beings and are literally linguistic in character. Whether it is your shoes or a walking stick, a zipper or a bulldozer, all of these forms are linguistic in structure and are outerings or utterings of man's own being. They have their own syntax and grammar as much as any verbal form. This was an unexpected result of looking at these innovations structurally, not with an intent to discover anything except individual structures. Eventually I realized that these structures are literally linguistic; there is no difference between hardware and software, between verbal and non-verbal technology in terms of this linguistic character or sharing.

This suggests, therefore, that man's technology is the most human thing about him. Our hardware – mechanisms of all types: spectacles, microphones, paper, shoes – all of these forms are utterly verbal and linguistic in character and are utterly human. The word *utter* is from *outer*, and "outering" is the nature of technology. Extension of bodily organs into the environment is a form of utterance or expression. There is, therefore, a completely intelligible character and pattern in these "outerings" or "utterings."

In an essay by Martin Heidegger titled "The Origin of the Work of Art," there is a wonderful passage in which he is talking about a pair of peasant shoes painted by Van Gogh.

We shall choose a well-known painting by Van Gogh, who painted such shoes several times. But what is there to see here? Everyone knows what shoes consist of. If they are not wooden or bast shoes, there will be leather soles and uppers, joined together by thread and nails. Such gear serves to clothe the feet. Depending on the use to which the shoes are to be put, whether for work in the field or for dancing, matter and form will differ.

Such statements, no doubt correct, only explicate what we already know. The equipmental quality of equipment consists in its usefulness. But what about this usefulness itself? In conceiving it, do we already conceive along with it the equipmental character of equipment? In order to succeed in doing this, must we not look out for useful equipment in its use? The peasant woman wears her shoes in the field. Only here are they what they are. They are all the more genuinely so, the less the peasant woman thinks about the shoes while she is at work, or looks at them at all, or is even aware of them. She stands and walks in them. That is how shoes actually serve. It is in this process of the use of equipment that we must actually encounter the character of equipment.

As long as we only imagine a pair of shoes in general, or simply look at the empty, unused shoes as they merely stand there in the picture, we shall never discover what the equipmental being of the equipment in truth is. From Van Gogh's painting we cannot even tell where these shoes stand. There is nothing surrounding this pair of peasant shoes in or to which they might belong – only an undefined space.

I would call that space an acoustic, resonating space.

There are not even clods of soil from the field or the field-path sticking to them, which would at least hint at their use. A pair of peasant shoes and nothing more. And yet –

From the dark opening of the worn insides of the shoes the toilsome tread of the worker stares forth. In the stiffly rugged heaviness of the shoes there is the accumulated tenacity of her slow trudge through the far-spreading and ever-uniform furrows of the field swept by a raw wind. On the leather lie the dampness and richness of the soil. Under the soles slides the loneliness of the field-path as evening falls. In the shoes vibrates the silent call of the earth, its quiet gift of the ripening grain and its un-explained self-refusal in the fallow desolation of the wintry field. This equipment is pervaded by uncomplaining worry as to the certainty of bread, the wordless joy of having once more withstood want, the trembling before the impending childbed and shivering at the surrounding menace of death. This equipment belongs to the *earth*, and it is protected in the *world* of the peasant woman. From out of this protected belonging the equipment itself rises to its resting-within-itself.

But perhaps it is only in the picture that we notice all this about the shoes. The peasant woman, on the other hand, simply wears them. If only this simple wearing were so simple. When she takes off her shoes late in the evening, in deep but healthy fatigue, and reaches out for them again in the still dim dawn, or passes them by on the day of rest, she knows all this without noticing or reflecting. The equipmental being of the equipment consists indeed in its usefulness. But this usefulness itself rests in the abundance of an essential Being of the equipment. We call it reliability. By virtue of this reliability the peasant woman is made privy to the silent call of the earth; by virtue of the reliability of the equipment she is sure of her world.[5]

Now the artist translates the hardware or equipment into another mode for contemplation.

Another idea that I would like to develop here is that of organized ignorance as an untouched resource. We are always trying to translate ignorance into knowledge and flip it on its back, as it were. But perhaps there is a way to organize human ignorance as a positive resource in this day of the mass audience created by electric media. The word *mass* simply means simultaneous. Mass man is man existing simultaneously in the same world. It is a matter of speed, not of numbers. It doesn't matter if it's six, or six million. If it is simultaneous, it is mass. This is part of Einstein's theories.

Now, suppose that we put questions to that mass through electronic media concerning the problems of our time; suppose that top researchers in various fields, in biology and chemistry and physics and town planning and so on, ask these questions; suppose they were to go to the broadcasting studios and present not their knowledge but their hang-ups; suppose they were to tell the mass, in the most succinct, atavistic, and structured form, where the difficulties are. Out in the mass audience, every single possible mode of perception exists unawares. But how do we tap that resource? I suggest that one possibility would be to take these highly specialist problems to this mass of untutored, non-specialist people. There is always one man in a million for whom any problem is not a problem at all. For a long time mathematicians used to pose as unanswerable the problem, "How far can you go into a forest?" One day some child simply said, "Halfway." And that is the answer; after that, you are coming out.

And it was an eight-year-old child who invented the cybernetic mechanism, called the governor, on a steam engine. It happened this way. He worked on a steamboat, and it was his job to pull the steamcock with a string. As the wheel went around, so the big piston went around. But he wanted to play marbles, so he tied the string to the flywheel and invented the first automatic governing mechanism in the world.

I suggest you will find historically that the greatest inventions were made anonymously by nobodies for whom there was no problem; they simply used common sense. All problems, when solved, in retrospect seem to have been easy. Why are they so hard when we are looking ahead at them? I suggest that to the untutored mass out there, all of the existing problems of the specialist are not problems at all. That is to say, what constitutes opacity when looking ahead is the misdirected knowledge of the specialist which shines in the face of the quester or pursuer. When a flashlight shines in your face you cannot see a thing. Now the specialist does that all the time; the flashlight of his specialty shines in his face, obscuring the answer to his problem. But there is always one man in a million for whom there is no problem. This feature of organized ignorance is a typical flip of the situation in which we live and I am going to note a few other situations which are similar.

At jet speed there is no rear-view mirror. What does one see in the rear-view mirror at motor-car speed? In the jet plane at jet speed, there is no rear-view mirror and nothing can be seen. What do you see in the rear-view mirror of a motor car? The foreseeable future. You don't see what went past, you see what is coming. It is obvious, isn't it? The phrase "rear-view mirror" tells you that you are looking at something that went past, but, in fact, you never do. All you can look at in the rear-view mirror is literally the foreseeable future.

Now, at the speed of light, there is no foreseeable future. You are there literally. It does not matter what situation you choose to consider. There is literally no possible future. You are already there the moment you name the situation. That is why in our age there are no goals. That is the reason for the streakers' antics: they are protesting the disappearance of goals. Where are we going? We are all dressed up with no place to go. We think we have taken all the right school courses, studied the right subjects, but now it all seems pointless. Where do we go from here?

Literally, there are no goals at the speed of light, but there are roles. At the speed of light, instead of having a job or an objective, you have to determine for yourself a totally new pattern, a new function in the world. And your new function is that of role-playing, which consists in taking on a whole variety of jobs at once. An ordinary mother with a family has many jobs to perform. So does a farmer. We consider them to be role-players. A farmer doesn't have a job and a mother doesn't have a job. They have roles, very involved and complicated ones. Role-playing is the new electronic form of job-holding; it is replacing job-holding. It does require a kind of reallocation of energies. At the speed of light all the old hardware pattern recognition is useless. If you do not know the new pattern in instant replay, you will be unable to make any use of it.

Hieronymus Bosch has a series of paintings which are currently popular and are considered to be examples of psychological horror. Bosch simply took the images of the preceding time, the time just before his own, and pushed them physically or pictorially through the images of his own time. By taking these medieval icons and pushing them into the new pictorial space of the Renaissance, a point-of-view space, he created horror. We are now living in a world where all the old pictorial space and all the point-of-view space of visual man are being inundated by the icons and the corporate images of advertising and huge public services.

All these contemporary electric forms have many, many medieval characteristics. They are acoustic; they are not visual. As these acoustic forms go riding through the old visual forms of our establishment, we get a confrontation of horror. This world of interplay leads, I think, to the world of the vampire movies, to the world of horror films, and is a kind of catharsis for that discomfort created by the clash of worlds. The totally alien and incompatible character of the visual and the acoustic, which Bosch in the fifteenth century presented as horror, is presented in our time, too, as horror, but in the opposite way. The scenery is arranged in the opposite pattern.

Joseph Conrad's world possesses this kind of image. In *Heart of Darkness*, Conrad faces a situation in Africa in which a group of Europeans are trying to exploit the ivory trade, and at the same time trying to civilize the natives. This is a peculiar interplay. Mark Slade has written about it in a recent essay in *Mosaic* magazine, in which he says:

> Man lives in the flicker, Man lives in the flicker. This comment was coined by Joseph Conrad in *Heart of Darkness*. He wrote the story about the same time moving pictures were first beamed out of a synthetic heart of darkness. Soon these electric images gave man twenty-four flickers a second to live in. It is in the flicker that man seizes a meaning for himself. But for Conrad's character the flickers race out of control in a riot of sensory inputs. Moving images also offer a riot of flickers which tumble after one another in such abundance that it is often difficult to fix a meaning. The effect in its most extreme form is something like being tickled to death.[6]

Slade confronts this inner-outer problem in this passage:

> A common error, however, is to regard the civilizing process and the humanizing process as synonymous. A correlation between the two has yet to be demonstrated. (The first glance at the place was enough to let you see the flabby devil was running that show.) . . . To be forced to return to an examination of our own origins may not be such a bad thing.[7]

We have long supposed that civilization and humanizing are the same thing. Civilization is a product of literacy, and humanizing may or may not be. Slade further says:

What can be claimed for literature and civilization is that they are euphemisms for hope. But hope for a better existence, for self-knowledge and understanding, could only be realized, so it seemed to Western man, by severing his ties with the tangled undergrowth of myth and superstition. He erected a greasy pole, the alphabet, on which he hoped to shinny out of the swamps.[8]

The civilized man imagines that he is going to help the native by stripping off the native's world of myth and legend, ritual and superstition. The paradox is that in the electric age we ourselves are moving into, returning to, the acoustic world of simultaneous involvement and awareness, experiencing the surfacing of the subliminal life. When all things are simultaneous, that is, at the speed of light, the things that are ordinarily put aside into the subconscious simply come up into the conscious. This is the meaning of Freud's *Interpretation of Dreams*. The surfacing of man's subconscious came with the telegraph, the telephone, radio, and then TV and other electric media. It is impossible to sublimate or keep anything hidden at that speed. So we have to invent a new concept of civilization and humanizing in order to live at the speed of light. We had imagined that we could simply strip off the acoustic culture of these primitives or these natives in order to civilize them, but the same stripping off of our civilization is taking place at the same time, by means of our new electric technology. We are losing our civilization even faster than we stripped off the institutions of the Indians and the Africans.

Slade goes on to say, "Commonly known, too, is the fact that many societies recoiled from the very idea of being wrenched from their primitive reality."[9] This is one of the big protests coming out of Africa right now. Africans do not want to be civilized. They do not want to become private people.

They want to retain their corporate social institutions. And as Slade further notes:

> Usually it had to be taken from them at the cost of massive slaughter. To willingly place themselves above the mud which gave them food, to tear away their own mythic roots, to wipe out the only survival pattern they knew, constituted the worst conceivable betrayal of a sacred trust. The sacrifice of individual human beings was one thing, the sacrifice of life itself was another.[10]

Mark Slade is simply talking about this paradox that we experience in moving into the acoustic world or simultaneous electric world, in which the world of the alphabet and of one thing at a time is not easy to maintain. Civilized Western man developed because of an alphabet which gave him a non-acoustic grasp of his world. He translated the whole of Homer and the acoustic encyclopedia into visual form and developed a whole new set of equipment of analysis and rational drive, goals and patterns. Euclid's theories were invented and new kinds of visual space discovered at the same time. This visual man with his rational, aggressive drive for goals was invented by this visual alphabet.

The visual alphabet to which we all are still subjected under the name of literacy is totally incompatible with instantaneous-simultaneous images of electric time. That does not mean one is good and one is bad. It just means they have completely different characteristics, and the values that are secreted by or brought to us by either form are not necessarily compatible any more.

I am not offering any solutions. I think that once you know where the structure of the problem is, it may be possible to hit upon solutions. But it certainly is very difficult to find solutions without awareness of where the problem is. I previously mentioned the novel by E. M. Forster, *A Passage to India*, in which the confrontation between the highly visual Westerner Adela Quested and her Oriental guests or hosts takes place in the Marabar Caves. Quested almost loses her mind as her strong visual sensitivity or bias encounters the acoustic world of the ear and the corporate non-individual society of the East. This is the situation in which all of us in the Fourth World are now engaged.

---

1. Hans Hass, *The Human Animal* (London: Newton Abbot, 1972), p. 101.

2. Ibid., p. 103.

3. Albert T. Simeons, *Man's Presumptuous Brain* (New York: E. P. Dutton, 1962), p. 43.

4. Anthony Storr, *Human Aggression* (London: Penguin Press, 1968), p. 112.

5. Martin Heidegger, "The Origin of the Work of Art," in *Basic Writings*, ed. David Farrell Krell (London: Routledge & Kegan Paul, 1978), pp. 162-64.

6. Mark Slade, "The New Metamorphosis," *Mosaic* 8 (1975), p. 131.

7. Ibid., 135.

8. Ibid.

9. Ibid.

10. Ibid.

Afterword

by David Staines

In September 1966, Marshall McLuhan, a member of the Department of English of Saint Michael's College in the University of Toronto, entered a small, second-story classroom in Teefy Hall to give the introductory lecture in his fourth-year honors course on "Modern Poetry and Drama."

By this time McLuhan was already well known to all of us. His first three books, *The Mechanical Bride: Folklore of Industrial Man* (1951), *The Gutenberg Galaxy: The Making of Typographic Man* (1962), and *Understanding Media: The Extensions of Man* (1964), had made him the pre-eminent theoretician and critic on communications and the new media; people from the business world and the newspapers both within Canada and from abroad were constantly seeking his views and advice. But the professor who stood in front of us on that September morning was not the man we had anticipated.

McLuhan entered alone, without any fanfare, reminding me more of the well-dressed manager of a local bank than of the media visionary we knew he had become. His dress, as was his custom, consisted of a dark suit and tie. When he began to speak, the words came effortlessly, the ideas were engrossing, and our minds were being enlarged by his probing adventures.

McLuhan and David Staines.

Ours was a literature course, and McLuhan was completely comfortable in the realms of modern poetry and drama. Each lecture was a detailed exploration of one of the authors we were studying or of one of the major ideas behind modern culture. Often he drifted away from the announced subject of his lecture as his ideas about the new media seemed to invade his arguments about T. S. Eliot or Ezra Pound. And yet his insights into the electric media arose naturally from the works he was discussing. The course became a dramatic introduction to modern literature through the eyes of a professor vitally alive to the shaping forces of contemporary culture.

Literature was McLuhan's home, we were his guests, and he was an amusing as well as a brilliant host. Many lectures began with jokes he had heard, and while our appreciation of these stories did not always equal his obvious zest in relating them, we marveled at his endless addiction to comic tales.

"I'm a professor of literature; I teach books from morning till night," he explained to Tom Brokaw. And he further re-iterated this idea to Mike McManus: "My values are strongly centered in literacy, which I teach night and day." Literature was central to his vision of the world, and most of his ideas about media arose naturally from the works he studied by Mallarmé, Rimbaud, and other French symbolists, for example, and James Joyce, and, of course, Eliot and Pound.

More than thirty-five years later, I still remember vividly his lecture on Eliot's *The Waste Land*. After a brief introduction to the poem, he set out to explore the power of allusion. "April is the cruellest month," he intoned, and then asked us what poem Eliot was evoking. Following a few moments' silence, someone answered, "Chaucer's *The Canterbury Tales*." Pleased with this reply, McLuhan launched into a comparison of Chaucer's poem with Eliot's work, pausing at length to recognize the world of Chaucer's pilgrims and the godless world of Eliot, which had no place for a pilgrimage because its inhabitants did not recognize any god or any saints. And so Eliot was invoking Chaucer's poem, he continued, to create a simultaneous reference point throughout his poem to another world that had the religious figures *The Waste Land* could not embrace. "The human city in *The Waste Land* is desiccated and deprived by mechanical repetition, whereas Chaucer's pilgrims of eternity represent the city in a very different way; they create their city as they go along the highway."

"[B]reeding / Lilacs out of the dead land," he then intoned, and asked us what poem Eliot was evoking. "Walt Whitman's 'When Lilacs Last in the Dooryard Bloom'd,'" I replied nervously. And McLuhan proceeded to discuss Whitman's poem, pointing out that his elegy for the dead Abraham Lincoln ("The city of man had been given a great blow by Lincoln's death.") was a perfect counterpoint to *The Waste Land*, for here too was no place for political heroes, for figures who could rally their people around a great vision.

Eliot's poem was, therefore, a modern *Canterbury Tales* where there was no realm for pilgrims. And Whitman's was an elegy that *The Waste Land* could not even fathom. As many times as I have read and taught Eliot's poem since that morning, I have always remembered McLuhan's account of the poem's opening lines and Eliot's sense of the power of allusion.

To his students, McLuhan was available in his office for consultation, advice, or just a chat; he was a professor who was on campus every day. There were many events and activities McLuhan graced with his presence. There was, for example, his public lecture on T. S. Eliot in my fourth year. There was his omnipresence when Sheila Watson, McLuhan's former graduate student, came from the University of Alberta to give an address. And, as I learned from Sister St. John, a venerable retired professor of classics, McLuhan was extraordinarily generous to the religious of the College faculty.

The noon hour was sacred for McLuhan's private world, for religion was an integral part of his life. Each day he could be seen crossing the campus to attend 12:10 mass at Saint Basil's Church. Nothing would be allowed to interfere with his daily attendance at mass and communion. Visitors who were going to lunch with him always gathered at his office at 12:45, never knowing the real reason for this later lunch hour.

McLuhan ended our course by saying that he would remember us in his prayers. The statement was simple, direct, and absolutely sincere.

That summer of 1967, McLuhan set off for Fordham University for the year as Albert Schweitzer Professor of the Humanities, and I, a naive twenty-year-old, went off to Harvard University to pursue graduate studies in English literature. Our paths did not cross again for any extended period until 1974 when I met him in Toronto for lunch, again at 12:45 p.m., to invite him to Harvard to deliver a public address in a course on Canadian literature I would be introducing the following year. "Why should I go to Harvard?" he asked me, then regaled me with reasons why Harvard did not intrigue him. "Why should I go to Harvard?" he asked again, and then, with no logical answer possible, I found myself stating, "Because of me." And, to my surprise, he replied, "All right."

During his Harvard stay I experienced McLuhan again as the knowledgeable and probing professor. When he came to my class the day after his public lecture to talk with the students, he was informed and informative, answering questions with responses that often ended with further questions, both educating the students and learning from them. As he sat discovering, illuminating, arguing, and enjoying the interchange, he was most at home and most himself. He was still, after being both lionized and discarded by the media, a true innocent, not in the sense of being unknowing or ignorant, but in the sense of being unaware of the designs of some of those around him.

These were many occasions when I met Marshall for lunch or for dinner, now as his colleague and no longer his student. These were times when we talked at length, usually alone, sometimes with his wife, Corinne, sometimes with one or two others.

We met daily for one week in June 1976, for example, when Marshall was finalizing and polishing his Harvard essay, "Canada: The Borderline Case," for publication in the forthcoming collection, *The Canadian Imagination: Dimensions of a Literary Culture*. "After you went off to your luncheon meeting," Margaret Stewart, Marshall's secretary, humorously informed me when I returned, "Marshall told me that he had finally met Simon Legree." When I greeted Marshall with news of the Legree comparison, he said that, of course, he meant it as a compliment!

In March 1977, when I was home in Toronto, a Harvard student of mine, who was doing his undergraduate thesis on the poetry of E. J. Pratt, was working at the Victoria College library, and I asked Marshall to meet with us for lunch. The conversation was almost totally devoted to Pratt, and Marshall quizzed the student energetically to learn from him.

On my Boxing Day visit in 1977, Marshall was celebrating the publication of *The Canadian Imagination*. He was so proud of his contribution, the closing essay in the volume, that he asked me to drive him down at once to his office, where he presented me with a perfect, first-edition copy of *The Mechanical Bride*. The copy was inscribed, "For David Staines, a great Canadian, in friendship and esteem from Herbert Marshall McLuhan." "I had to sign it that way," he added, "because that was my name when I published this book."

The last time I saw Marshall was the summer of 1979. At a dinner party he and I attended, there were guests from the business and educational worlds, and much of the evening found Marshall responding to them, offering advice, and playing the role of media visionary. At the end of the evening, I was anxious to leave, exasperated at the other guests' easy acceptance of the mythic figure and disappointed at my own inability to connect with the man I was so fond of. Marshall wanted to walk me to my car, and although I declined, he was insistent. When we were outside, he confided to me, his left arm around my shoulders: "I want to thank you for being here this evening. It is always so much easier for me at this kind of evening when someone who knows me is also present. It makes the evening possible to endure." Still and always the innocent, caught up in a web not completely of his making.

Marshall McLuhan died in the early hours of December 31, 1980. His funeral took place in his parish church of Holy Rosary a few days later. For many of us in attendance, it was fitting that our final meeting with Marshall take place inside the church he knew so well.

Index